Twenty-First Century Military Innovation

Twenty-First Century Military Innovation

Technological, Organizational, and Strategic Change beyond Conventional War

Marcus Schulzke

University of Michigan Press

Ann Arbor

Contents

Digital materials related to this title can be found on the Fulcrum platform via the following citable URL:
https://doi.org/10.3998/mpub.11954694

Introduction

Twenty-First Century Military Innovation

Contemporary war is as much a quest for decisive technological, organizational, and doctrinal superiority *before* the fighting starts as it is an effort to destroy enemy militaries *during* battle. Armed forces that are not actively fighting are instead actively re-engineering themselves for success in the next fight and imagining what that next fight may look like. As the twenty-first century moves into its third decade, it seems certain that the speed of military advancement will continue undaunted and that many of the key decisions shaping future conflicts will be made away from the battlefield.

I argue that the direction of innovation and its broader political context have changed considerably since the end of the Cold War, and in ways that necessitate a different understanding of the broader political and military significance of military innovation. This is a time when new technological opportunities are being actualized not for the consolidation or expansion of the Westphalian system, but as a corollary of that system's decline.[1] As I show

1. The roots of the sovereign state, and the system of international politics between states, are generally traced to the Treaty of Westphalia in 1648. I will accordingly refer to this as the "Westphalian system" for the sake of simplicity, though this name is somewhat misleading because the system emerged gradually, rather than being immediately produced by the signing of a treaty. Westphalian states are characterized by a strong conception of sovereignty, according to which states have absolute authority within their own borders and a monopoly on the legitimate use of coercive force therein. The Westphalian system comprises states that collectively uphold sovereign authority, territorial claims, and legitimate authority to use lethal force, while denying those rights to nonstate entities. The Westphalian system took shape during the seventeenth and eighteenth centuries, but only became the dominant political logic during the nineteenth and twentieth centuries, or, to be more precise, between the Napoleonic Wars and the end of the Cold War. It was during that time that states participating in the Westphalian system used the opportunities afforded by industrialization to achieve new advantages over rival states and to assert dominance over political entities that lacked recogni-

right (such as with sanctions). Conceptions of nonmilitary power, including "soft power" and "public diplomacy," are increasingly popular in strategic discourse, but in practice these rely heavily on being paired with military force that can work as a coercive instrument underwriting persuasion.[5] War may not be reducible to politics, but it would be impossible to understand war or politics without attention to how these two practices constitute each other. Thus, the innovations discussed throughout the book are not confined to military affairs. They help to constitute politics and civil society.

I illustrate the ironic effects of strong states' pursuit of military advantage by exploring eight of the most radical ongoing shifts in military innovation beyond the domain of conventional war and by analyzing the deeper thematic currents that run through each. The eight topics are arranged into three sections to encompass general areas of change: military technologies (exemplified by drones, cyberweapons, and nonlethal weapons), the organization of armed forces (exemplified by private military contractors and special operations forces), and methods of fighting (exemplified by targeted killing, population-centric counterinsurgency, and degradation). These are by no means the only innovations affecting twenty-first-century warfare, yet they are among the most theoretically important. Moreover, treating them as case studies indicative of broader trends makes it possible to gain a sense of the direction of development in warfare and how domestic and international politics are changing along with it. The innovations I discuss share four characteristics: novelty, usability, diffusion, and disruption. The innovations are not meant to constitute a random sample. They were selected with the theory-building goal of showing how states and their armed forces are responding to, and fueling, a transformation of war beyond conventional interstate conflict.

Novelty

Novelty is the most obvious of the four selection criteria. Innovations must in some sense be new to qualify as innovations. Most of the innovations I discuss have been enabled by technologies that emerged in, or matured in, the past forty years. Several of the innovations have historical roots that can

5. Joseph Nye, originator of the concept of smart power, describes this as being partially built on strong alliances and relations of influence, which are often themselves created and sustained by military power. Moreover, the "smart power" that he favors combines soft power with hard military power. Joseph S. Nye, *Soft Power: The Means to Success in World Politics* (New York: Public Affairs, 2004).

be traced back centuries, but are novel insofar as they are coming back into conflicts after centuries of marginalization. Thus the innovations selected are either new in the sense that they depend on the latest technologies, as in the case of cyberwarfare, or in the sense that they show the reappearance of methods of fighting that the Westphalian system suppressed, as in the case of private military contractors. In neither of these senses does novelty mean that the innovations are wholly unprecedented. Innovations do not spring from the ether, and even the most ground-breaking embody abiding cornerstones of military thought.

The emphasis on novelty in the case studies shapes the overall scope of the analysis. I go beyond security issues, narrowly construed, to discuss how the innovations are shaped by interactions between military and civilian domains. For instance, I consider the ways in which hobbyist drones have incorporated military technologies and vice versa. I likewise devote some attention to the use of new media, such as video games, to train soldiers and to allow armed forces to shape narratives about ongoing and potential conflicts. This reinforces my contention that the boundaries of war are blurry, to the extent that many domestic activities can be seen as participating in conflicts to varying degrees.

Usability

The innovations selected for case studies are "useable weapons"[6] in the sense that they have a low bar for use in a broad range of conflict scenarios. The wars of the nineteenth and twentieth centuries showed a proclivity for overwhelming firepower that was primarily useful in conventional interstate wars. Participants in those wars created a trap for themselves. They made conventional wars destructive to the point of being untenable, and they were left with few military options for exerting low levels of coercion outside conventional warfare. Usable weapons are closely associated with states bending the rules of the Westphalian system, as these can be used in ways that are difficult to classify within the conventional war vs. peace binary. Moreover, because usable weapons are often deployed for domestic law enforcement and in unconventional wars against violent nonstate actors, their creation and implementation constitute a tacit acknowledgment that war cannot be confined to the conventional model.

6. David A. Koplow, *Death by Moderation: The U.S. Military's Quest for Useable Weapons* (New York: Cambridge University Press, 2010).

Military innovation continues outside this category of usable weapons, and other types of weapons remain central to military strategy. My focus on usable weapons is not meant to downplay more destructive instruments of conventional war but instead to show how armed forces are extending their capacities to provide a more comprehensive response to threats. In fact, the continued investment in extremely destructive conventional weapons is one of the key reasons why states have invested in more usable tools of coercion. Nuclear deterrence has been especially important in prompting states to develop coercive instruments that limit the likelihood of escalation. As the potential costs of interstate war increase, so too do the potential gains of developing usable weapons that enable alternative methods of fighting.

Diffusion

Many prospective weapons systems never leave the development stage. They are designed, built, tested, and perhaps selectively used, without having a significant impact on the conduct of war. The diffusion criterion is meant to exclude such innovations. Each of the innovations I discuss shows diffusion in three different ways. First, there is diffusion within a particular military, such that the innovation is used across multiple branches of service and often beyond the role it was initially meant to serve. Second, there is diffusion to other countries. Widespread adoption is evidence that an innovation is likely to have lasting importance and that it is likely to play a role in future conflicts, regardless of the participants. Third, there is diffusion between military and domestic settings. One of the central arguments running throughout the book is that the boundary between war and law enforcement is blurring and that this process is fueled by the sharing of technologies and practices between militaries and law enforcement agencies.

Disruption

The innovations I discuss all have disruptive effects on existing norms governing the use of force. They do not fit established notions of conventional warfare, nor do they fit Westphalian notions of the sovereign state. They work beyond those boundaries to create new contexts for violent political action, in which states are freer to exert selective coercion. This positions the innovations to have a greater impact beyond the military than weapons conforming more with Westphalian expectations of conflict being confined to a particular domain. The disruption is primarily evident from the challenges the innovations present to the just war tradition and the professional military culture. Those two normative systems have acted as a kind of informal rule book for conflicts between sovereign states, with the former providing guidelines for how wars should be conducted and the latter providing guidelines for how armed forces should be composed and what values military personnel should internalize. I discuss the details of these normative systems throughout the book as I call attention to the ways they are being disrupted.

When its norms are disrupted, the Westphalian system loses some of its legitimacy, and the Westphalian conception of conventional war changes. This creates a dilemma. The norms can be patched up and preserved to accommodate the innovations, which would inevitably make the norms more contradictory over time, or the norms can be thoroughly revised via a decisive break with the past. So far, the former approach is winning, with the result that the norms governing war are slowly breaking down or being applied opportunistically.

Overview of the Cases

The three technological innovations I discuss are armed drones, cyberattacks, and nonlethal weapons. Each is new in the sense of recently emerging and depending on underlying technologies that are themselves of recent origin. They are highly usable because they are designed for a broad range of conflict scenarios ranging from major conventional wars to intelligence collection and policing. They are spreading rapidly, not only across countries but also beyond normal military spaces. In doing so, these innovations have prompted sustained debates about whether they are morally and legally justifiable.

Of the three technologies I discuss, drones are having the clearest and most immediate impact on the conduct of war. Unmanned aerial vehicles (UAVs) make it possible for attackers to launch strikes against enemies from extreme distances in what is often characterized as "riskless war." It has become common for American, British, Russian, and Israeli pilots stationed within their home territories to monitor and attack suspected terrorists and insurgents without exposing themselves to any danger of reciprocal attack.

extent of diffusion, with most participants relying heavily on military contractors and special operations forces to perform key missions.

The widespread appeal of these two organizational innovations has much to do with their heightened usability compared to conventional forces. The extensive training of these personnel makes them highly adaptable and able to exert selective force to satisfy capricious political necessities. Both groups are disruptive insofar as they blur the boundary between combatants and civilians, change the nature of accountability for wars, and challenge the traditional rationales for participating in wars.[14] It is important to note that there is a reason why I limit the analysis to two organizational case studies in contrast to the three case studies each for technological and strategic change. Armed forces tend to be more open to innovation in technology and strategy, while being comparatively less receptive to organizational restructuring. Major organizational changes are correspondingly less frequent and, as I will show in the case studies, tend to occur in large part because of the necessities of implementing innovations in the other two domains.

Private military contractors follow in a long tradition of combatants fighting for pay. Western armed forces relied heavily on mercenaries during the Middle Ages and Renaissance,[15] before states consolidated power and established the monopoly on coercive force that is one of the defining features of the Westphalian system. The re-emergence of privatized armed forces indicates declining state control over war, and a corresponding threat to the Westphalian system. Weak states turn to contractors to offset institutional limitations that prevent them from maintaining conventional armed forces. They struggle to protect their status as well-ordered members of the international community, but ironically they do this by empowering an industry that threatens the ideal of sovereign authority. Strong states may be capable of defending themselves against regular adversaries, yet they resort to private assistance to cope with the new responsibilities of counterinsurgency and peacekeeping operations that have exposed gaps in existing military

---

14. These changes are not entirely new; the line dividing combatants from civilians has shifted considerably over time, and accountability mechanisms are continually changing along with governmental institutions. Whereas the processes of formalizing civilian/combatant distinctions and professionalizing armed forces during the nineteenth century cohered with the prevailing norms and efforts to formalize international law, current changes in the composition of armed forces are generally antagonistic to existing norms of war and political institutions.

15. Sarah Percy, *Mercenaries: The History of a Norm in International Relations* (New York: Oxford University Press, 2007); Thomas Arnold, *The Renaissance at War* (London: Cassell, 2001).

expertise. Contractors take on work like protecting commercial shipping from pirates, guarding civilian government officials, and giving logistical support to armed forces. Here they show that even ostensibly strong states are struggling to adapt to emerging security challenges and that the quick fix of privatized assistance is often more attractive than the painful process of fundamentally restructuring armed forces to meet emerging threats. This dependence raises serious concerns for all states, regardless of their domestic military capacities. Whereas mercenaries were once part of the accepted order of war, private contractors now occupy a gray area. They stretch the boundaries of just war theory and international law that have been built to fit the Westphalian model. Moreover, contractors challenge the sense of civic obligation that has been linked to military service, threatening to bring about a collapse of the nationalistic image of war as a morally obligatory and praiseworthy act of self-sacrifice.

Private military contractors have been widely scrutinized and heavily criticized; the organizational changes caused by the growing reliance on special operations forces tend to receive less attention. Here the changes are less overt, but equally significant. More soldiers than ever are needed for noncombat support roles.[16] Military specializations are becoming narrower. Fewer military personnel are involved in traditional forms of combat, as evidenced by the three technologies I discuss. Military tactics and culture are changing as more soldiers in regular units are asked to do what is essentially police work. And states face greater demands to restrict the number of participants in combat to minimize the risks of sustaining or inflicting casualties. Special operations forces have the potential to maintain a high degree of coercive power even with fewer personnel being committed in ground combat. They also embody the military culture that is being disrupted by new technologies, helping to sustain a sense of national pride in service and affirming the value of self-sacrifice. In many ways, they are the perfect antidote to the ostensible moral degradation of contractors and insurgents, as well as the technological rationalization of drones, cyberwarfare, and nonlethal weapons. They are understandably attractive to policymakers aiming to wage wars with minimal risk. However, like private military contractors, special operations forces indicate growing inequality on the battlefield. When small numbers of combatants play a disproportionate role in the fighting, war loses the ethos of mass participation that has animated it since the Napoleonic Wars and takes on a post-Westphalian character.

---

16. John McGrath, *Other End of the Spear: The Tooth-to-Tail Ratio in Modern Military Operations* (Fort Leavenworth, KS: Combat Studies Institute Press, 2011).

The final part of the book covers three of the distinctive strategic/tactical changes in twenty-first-century warfare: targeted killing, population-centric counterinsurgency, and degradation. These are best conceptualized as being joint strategic/tactical changes because much of their disruptive impact comes from how they undermine this distinction and make the two concepts difficult to extricate.[17] The three innovations in this domain are novel because they depend on emerging technologies, including the three weapons technologies I discuss. They enhance the usability of coercive tools by extending violence beyond conventional warfare to such an extent that they make it difficult to identify a line between war and peace. Finally, these practices have undergone considerable diffusion between states attempting to find the most effective methods of employing emerging weapons systems.

Targeted killing epitomizes some of the most important processes underlying the merger of tactical and strategic thinking, as well as the absence of a clear dividing line between war and peace. It embodies an effort to reduce battlefield casualties to the absolute minimum, to the point that it is difficult to say whether operations qualify as war, law enforcement, or something else entirely. This is made possible by abandoning the goal of inflicting massive casualties on rank-and-file combatants, which is at the heart of conventional warfare. Leaders, vocal ideologues with prominent media platforms, and support personnel are the new focus, and operations are guided as much by political calculations as by battlefield objectives. Moreover, targeted strikes are orchestrated by complex assemblages of civilian and military actors. This is partly because operations lend themselves to a high degree of centralization and partly because they depend heavily on intelligence collection, diplomatic, legal, and military processes. Targeted killings have become one of the most popular tools for US, UK, and Israeli policymakers who struggle with

17. "Tactics" usually refers to the practices that small and mid-sized units employ to defeat their opponents on the battlefield, while "strategy" refers to the higher-level decisions about how to maneuver units and when to give battle. In a sense, tactics, as well as operational matters between tactics and strategy, fall under the purview of military decision makers, while strategy is more heavily politicized. The logic of these three innovations leads intelligence agencies and politicians to participate more directly in affairs on the ground. Conversely, soldiers operating in the tactical domain take on profound strategic and political importance. The three innovations even upset the more fundamental distinction between war and politics, with each depending on an unprecedented merger of military and political functions and a streamlining of the interactions between them. These may therefore be understood as new techniques of war fighting that transcend the conceptual distinctions that once helped to bring some sense of order to war. With this in mind I sometimes speak of "strategy" in a loose sense by using this term to encompass actions that many authors disaggregate along tactical, operational, strategic, and political levels.

Theme 2: Mediation

Mediation is the process of filtering information or actions through one or more intermediary layers.[27] When applied to the innovations I discuss, it comes in two distinct forms: physical and institutional, which I group together because they are functionally similar and often found together. Physical mediation is a matter of using space or some kind of material to diffuse adverse effects. For example, extending the range of weapons to limit an adversary's opportunities for retaliation. Institutional mediation is a matter of adding layers into decision-making processes. For example, if control of the long-range weapon passes through a complex bureaucracy that shares decision-making functions.

Physical mediation changes the experience of war in ways that break conventional war relationships of reciprocity and symmetrical exposure to risk. Before the twenty-first century, war usually required direct physical confrontation between opponents. It was only mediated in a limited sense by short-range weapons like swords, spears, and bows that extended combatants' reach. Gunpowder weapons increased mediation, first by detaching fighting ability from physical strength, then by increasing the range of conflicts to the point that enemies could be engaged without actually seeing each other. Soldiers came under fire from artillery, bombs, armored vehicles, and long-range small arms. Violence emanated from people who were largely invisible. Contact with the effects of enemy weapons, rather than direct confrontation with the enemy, was the predominant experience of combat. However, combatants on opposing sides were equally affected by this experience of anonymous warfare, which provided the basis for a shared perspective on war. In the twenty-first century, virtually every novel weapon system and organizational change (including advancements in conventional warfare that I do not consider here) is designed to place real or figurative layers between opposing combatants, whether it is through computerized targeting systems or proxy security forces. This further distances combatants from each other to such an extent that those on opposing sides often have much different experiences of war.

Physical mediation is significant for audiences watching from a distance because it alters the form and content of information in ways that have mixed

27. Note that this meaning of the term "mediation" is different from mediation as a process of dispute resolution through nonviolent means. Both senses of "mediation" come up in security contexts, though I exclusively use the term as I define it here, and not to describe arbitration.

positive and negative implications. At the most fundamental level, mediation extends our view of the world. "Mediation brings what is beyond our human sense experience. It brings us representations of that which is beyond our immediate surroundings."[28] Physical mediation through digital communications devices alters the experience of war for the public watching from a distance. For the first time in history, it is possible for audiences around the world to watch gunfights and missile strikes. They gain a sense of direct access to combat. However, careful management of mediated experiences creates new opportunities for controlling this seemingly authentic information and fitting it into self-serving narratives.[29]

Institutional mediation arises from the challenges inherent in controlling complex weapons and organizations. Generally speaking, the more advanced armed forces become, the more they depend on an assortment of specialized skills that can only be found by including multiple people and agencies in decision-making processes. This type of mediation aids the management of political risks by making it easier for government officials to avoid oversight and shift blame. Blame shifting is a well-known feature of bureaucracies,[30] but institutional mediation takes on a unique character in the twenty-first century as new technologies, organizational structures, and tactical/strategic objectives facilitate this process by expanding the range of options available to those attempting to manage perceptions.

Physical and institutional mediation are complementary. When wars are waged without direct contact with the enemy (via physical mediation), it is easier to fight without being clearly linked to specific decisions or outcomes (via institutional mediation). Mediation has rendered conflicts more asymmetric and has introduced new capacities for engineering representations of war for audiences and participants alike. Mediation generally increases belligerents' capacities for managing risks. Employing weapons that produce physical mediation, such as cruise missiles and drones, not only creates a dis-

28. Andrew Hoskins and Ben O'Loughlin, *War and Media* (Cambridge: Polity Press, 2010), 63.

29. For examples of how media can be manipulated to present self-serving views of conflict for audiences at home, see James Der Derian, *Virtuous War: Mapping the Military-Industrial-Media-Entertainment Network* (New York: Routledge, 2009); Susan L. Carruthers, *The Media at War* (New York: Palgrave Macmillan, 2011).

30. Dennis Thompson, "Moral Responsibility of Public Officials: The Problem of Many Hands," *American Political Science Review* 74, no. 4 (1980): 905–16; Christopher Hood, *The Blame Game: Spin, Bureaucracy, and Self-Preservation in Government* (Princeton, NJ: Princeton University Press, 2010).

tinctive experience of fighting but also allows belligerents to minimize risk to their own personnel, who are often located in safe areas hundreds of miles away from the target. Increasing the accuracy of these weapons makes it possible to assert greater control over which specific people in the target area will be in danger. Perceptual risks, such as those associated with public backlash against the handling of conflicts, are also minimized when mediation is used to opportunistically shift responsibility.

Theme 3: Perception Management

In democratic states, accountability depends on the public's being able to identify the parties responsible for the conduct of military affairs and to reward or punish their decisions.[31] The public is supposed to act as a check on reckless or improper uses of force by removing politicians from office, but the public response to conflicts and use of accountability mechanisms depends on what information members of the public receive and how people understand conflicts they usually only experience from a distance, through layers of mediation.[32] Perception management is a key consideration for civilian and military leaders who must pursue their agendas while maintaining public support in a world that is increasingly hostile to instrumental political violence. Internationally, managing perceptions is essential for maintaining alliances, avoiding punitive actions from other states, and convincing partners to undertake multilateral military ventures. Creating the right impression may even be a primary goal during war, as in the case of population-centric counterinsurgency operations in which indigenous support is essential to long-term security. The 2014 US counterinsurgency manual says counterinsurgency forces should "carefully consider that information reactions impact on the many audiences involved in the conflict and on the sidelines. They should work actively to shape responses that further their ends."[33] Throughout the book I show that this thinking goes far beyond counterinsurgency to shape conflicts of virtually any type, includ-

31. Shanto Iyengar, *Is Anyone Responsible? How Television Frames Political Issues* (Chicago: University of Chicago Press, 1994); Lonna Rae Atkeson and Cherie D. Maestas, *Catastrophic Politics: How Extraordinary Events Redefine Perceptions of Government* (Cambridge: Cambridge University Press, 2012); Hood, *The Blame Game.*

32. James Igoe Walsh and Marcus Schulzke, *Drones and Support for the Use of Force* (Ann Arbor: University of Michigan Press, 2018).

33. *FM 3–24/MCWP 3–33/5 Insurgencies and Countering Insurgencies* (Washington, DC: Department of the Army, 2014), 1–140.

ing those between states. Russia in particular has been adept at using disinformation campaigns to weaken its enemies.[34]

Koplow is insightful in noting that there is an ethical component to perception management. "Image also plays an important role in moderating our conquests—both self-image and preservation of the portrait of us that we want others to hold. Our sensibilities dictate that we do not want to 'look bad,' even when we are performing the dirtiest, most gruesome tasks of combat."[35] In this sense, perception is tied to the concept of legitimacy, which is the "generalized perception or assumption that the actions of an entity are desirable, proper, or appropriate within some socially constructed system of norms, values, beliefs, and definitions."[36] Actors have an interest in maintaining their legitimacy as a component of their ability to coerce or persuade others, including their own fighters. Legitimacy can also act as a weapon when attempts are made to discredit adversaries through misinformation campaigns.

Perception-management practices are further evidence of the importance of risk management and mediation. Because risk evaluations are premised on imagined futures, the choice of what information is fed into risk calculations is extremely important. Factual inaccuracies and malformed beliefs are potential security vulnerabilities. Mediation provides the kind of information control that makes it possible to perpetuate self-serving narratives and to deny opponents a platform. Virilio aptly notes that "since the beginning of the [twentieth] century, the news has constituted a fourth front"[37] of war because of this capacity for controlling meaning. This insight has become more salient over time as the extent of media coverage has increased, the costs of production have declined, and digitization has accelerated the circulation of content. Now state military forces and violent nonstate actors alike pursue sophisticated media strategies that often involve the creation of their own persuasive material. Armed forces have taken an active role in sponsoring or producing entertainment that can help to advance strategic communications goals and facilitate the quest for legitimacy.[38] Ostensibly neutral

34. Alexander Lanoszka, "Russian Hybrid Warfare and Extended Deterrence in Eastern Europe," *International Affairs* 92, no. 1 (2016): 175–95.

35. Koplow, *Death by Moderation*, 42.

36. Mark Suchman, "Managing Legitimacy: Strategic and Institutional Approaches," *Academy of Management Review* 20 (1995): 571–611, 574.

37. Paul Virilio, *Desert Screen: War at the Speed of Light* (New York: Continuum, 2002), 67.

38. Roger Stahl, *Militainment, Inc.: War, Media, and Popular Culture* (New York: Routledge, 2009); Roberto J. Gonzales, *Militarizing Culture: Essays on the Warfare State* (Walnut

news agencies and entertainment producers are likewise brought into the war effort with incentives like production assistance and greater access.

Theme 4: Disruption of Geography

War in the twenty-first century transforms geography, at least as it is understood in strategic and political thought shaped by Westphalian understandings of sovereignty and conventional war. In the past, fighting was necessarily grounded in a contested space and usually motivated by the goal of controlling that space. War was conceptualized in terms of territorial intrusion; one polity had cause for war when another encroached on its territory. The goal was usually control of the local resources, whether these were extractible resources or the land itself. The geographical conception of war was formalized under the Westphalian system, with states receiving absolute authority to resolve conflicts within their own borders and having cause for war against any states violating those borders. The concept of sovereignty that emerged during this time hinged on the distinction between a state's internal monopoly on force and the external anarchy of the interstate system.[39] That link between war and geographical space is attenuating. Conflicts are now routinely waged for reasons other than the traditional goals of asserting territorial control, punishing territorial infractions, and securing natural resources. And operations are not restricted to a determinate battlefield. Moreover, the Westphalian conception of sovereignty is questionable in an era when states breach the integrity of other states by declaring them to be "failed" or "rogue" states beyond the Westphalian norms.

Some commentators have attempted to show that war is necessarily grounded in geography, regardless of the prevailing political system. For example, Rachel Woodward argues that "Armed conflict, as we all know, is inherently spatial in its expression and constitution."[40] I do not want to suggest that geography no longer matters but rather that it does not matter in the same ways it once did. Wars continue to take place within geographic spaces, and even cyberattacks must have physical components, yet the reasons for fighting have changed considerably to less tangible goals like building legitimacy, creating new governments, and managing risks. Although war retains physical

---

Creek, CA: Left Coast Press, 2010).

39. David Parrott, *The Business of War: Military Enterprise and Military Revolution in Early Modern Europe* (New York: Cambridge University Press, 2012), 264–67.

40. Rachel Woodward, "From Military Geography to Militarism's Geographies: Disciplinary Engagements with the Geographies of Militarism and Military Activities." *Progress in Geography* 29, no. 6 (2005): 725.

dimensions, state military forces tend to measure their advancement in terms of how readily they can supersede geography by launching attacks that do not involve ground forces, deploying highly mobile forces, and employing strategies that allow them to avoid directly competing for territorial control. There are no longer distinct spaces of war apart from spaces of peace. Unlike their predecessors, twenty-first-century armed forces can fight virtually anywhere and aspire to full-spectrum dominance. They do not fight on bounded battlefields but within potentially unlimited battlespaces.

The geography of conflict is also changing in the sense that ground combat is increasingly urbanized.[41] "The economic, social, and political character of cities makes urban areas lucrative targets for a force wanting to control or influence a nation."[42] Whereas armed forces once avoided fighting in urban areas, preferring open country where they could maneuver more easily, they are now compelled to conduct their most important operations in built-up areas.[43] The war of maneuver that was the pinnacle of conventional operations is no longer the key to victory; more often than not, the key is to win a war of occupation. And in another twist on traditional understandings of geography, control of these spaces is less about the seizure of territory than the seizure of populations. The "human terrain" takes precedence over physical terrain. Legitimacy, which is an aspect of perception management, is as important as possessing land. Fighting in urban areas likewise entails countless new challenges associated with interacting with civilians and entrusting armed forces to participate in governance, challenges that alter the nature of war by making it more like law enforcement. The impetus toward technologies and tactics that can restrict violence, as well as organizations that can apply it more precisely than conventional militaries, is heavily influenced by the new priorities introduced by urban combat.

Theme 5: Individuation

Individuation refers to the increasing salience of individuals as compared to collectives. Conflict is becoming more heavily individuated in three senses. First, people are participating in conflicts more as individuals who have

41. David Kilcullen, *Out of the Mountains: The Coming Age of the Urban Guerrilla* (New York: Oxford University Press, 2013).

42. Glenn J. Gray, *The Warriors: Reflections on Men in Battle* (New York: Harper & Row, 1967), vii.

43. Anthony James Joes, *Urban Guerrilla Warfare* (Lexington: University Press of Kentucky, 2007); Stephen Graham, *Cities Under Siege: The New Military Urbanism* (New York: Verso, 2010); Kilcullen, *Out of the Mountains*.

their own interests and autonomy intact. Conventional war has been understood as a contest between opposing groups, in which fighting occurred because of intergroup enmity. Individuals had limited scope for independent action and were supposed to fight on behalf of collective goals they might feel little personal stake in. Although the nature of the groups involved in warfare has changed substantially over time (for example, wars involved states, tribes, empires, etc.), a collectivist view of war has been relatively consistent. This conception of war reached its apogee under the Westphalian system, which made war a matter of state "fought exclusively by sovereigns as a just means of the acquisition of property"[44] that had to be waged for collective rather than narrow personal interests. Violence perpetrated by other entities was classified as criminal or deviant. Even heads of state could not justifiably initiate hostilities for personal motives and had to ground violence in other (usually territorial) rationales. This is exemplified by Jean-Jacques Rousseau's discussion of the political relationships that give rise to war:

> War is not, therefore, a relationship between man and man, but between state and state, in which individuals become enemies only by accident, not as men, nor even as citizens, but as soldiers; not even as members of their own nation, but as its defenders. Furthermore each state can be enemy only to other states, and not to men, given that between things diverse in nature no true relationship can be established.[45]

Twenty-first-century conflict stands in stark opposition to this sentiment. It reveals growing emphasis on individual contributions and a rejection of anonymous mass warfare. Armed forces are increasingly compelled to devolve authority to lower levels, which opens a space for combatants to make strategically and politically important decisions more independently than in the past. They are likewise reconceptualizing war as a practice in which individual participants need their own goals apart from those of intergroup enmity. To update Rousseau, war is now not only a relation of state to state but also one of man to man and group to group. States may have individuals as their enemies, and vice versa.

---

44. Stephen Neff, *War and the Law of Nations: A General History* (Cambridge: Cambridge University Press, 2005), 19.

45. Jean-Jacques Rousseau, *Discourse on Political Economy and the Social Contract* (Oxford: Oxford University Press, 1994), 51.

Second, a more restricted range of actors is involved in any particular type of fighting as military roles become more specialized and technically demanding. Small units of special operations forces and drone operators track down and kill high-profile terrorists, with only a handful of people on each side participating in violence that could alter the course of entire wars. Even the burden of conventional operations falls disproportionately on a limited number of personnel.[46] Differentiation of roles naturally produces enclaves of expertise that imbue insiders with more power and more opportunities to resist outside pressure. Participation is also narrowing when it comes to who is classified as an enemy. The demands of managing risks and perceptions push belligerents toward narrowly focusing violence against specific targets, sometimes people who are known by name. At the turn of the twentieth century military innovation was geared toward total annihilation of anonymous enemies; now it is guided by the goal of focusing violence so narrowly that an attack can kill a single person in a crowd.

Finally, individuals have come to play a major role in framing conflicts, to a large extent because individuals act as gatekeepers in processes of mediation. Wars are often explained to audiences in terms of individual stories, and exceptional individuals have become "strategic corporals" who take on importance far beyond their ranks and responsibilities. The strategic impact is especially pronounced when these soldiers do things that garner media coverage. This is evidenced by the infamous soldiers who participated in the Abu Ghraib prisoner abuse. The pervasiveness of media means that virtually any soldier could potentially become a celebrity combatant, taking on more symbolic importance than the anonymous "boots on the ground," who fade into relative obscurity. On the other hand, celebrity figures on the opposing side have become prime targets. Those involved in perception management—propagandists and clerics like Jihadi John and Anwar al-Awlaki—have nearly as much significance as enemy commanders.

Theme 6: The Changing Military Profession

The military is a profession in the sense that it "is aware of its special and separate identity distinguishing it from civilian corporations."[47] It maintains its own culture, norms, institutions, and practices, self-reflectively guarding these against outside influences. Samuel Huntington identifies

46. McGrath, *The Other End of the Spear*.

47. S. E. Finer, *The Man on Horseback: The Role of the Military in Politics* (Boulder, CO: Westview Press, 1988), 54.

three characteristics of professions: expertise, responsibility, and corporateness.[48] The military profession that was created to conduct conventional warfare is being disrupted when judged according to each of these criteria. Soldiers are no longer the sole masters of coercive instruments, or even the ones who are in the best position to use them. Armed forces increasingly depend on technicians to build and operate advanced weapons, contractors to provide logistical support, law enforcement officers to assist in low-intensity operations, and intelligence operatives to direct attacks. This makes it impossible for armed forces and other state security forces to maintain the Westphalian monopoly on coercive force. Military institutions and cultures are also profoundly disrupted by new weapons and methods of fighting. For example, increasingly mediated weapons like drones give war a much less romantic appearance, and counterinsurgency makes soldiers act more like police officers. The changes I discuss simultaneously strike at expertise (by demonstrating that the military no longer has all the requisite skills for fighting and that other institutions may be more suited for certain roles), responsibility (by clouding the links between security providers and eroding oversight mechanisms), and corporateness (by transforming the military from a coherent institution to one with permeable boundaries and difficulties delineating membership).

The concept of a military profession often relies on the assumption of a primordial warrior identity that transcends time and place. Soldiers still look back to figures of antiquity like Achilles and Hector as reference points.[49] However, by most accounts, military professionalism is a recent phenomenon, beginning to take shape during the late nineteenth century and only really reaching its height during the mid-twentieth century.[50] Given the profession's novelty, it is evident that professionalization is not essential to being a warrior and that war is possible without professionalism. The dissolution of the profession is upsetting efforts to reify it as something fixed and timeless. This in turn threatens notions of sovereignty that depend on states monopolizing the legitimate use of coercive force within a particular territory via professional armies made up of citizens. Changes to the military profession also disrupt the political relationship between soldiers and the state, as a loss of professionalism creates opportunities for the politicization of armed forces.

---

48. Samuel P. Huntington, *The Soldier and the State: The Theory and Politics of Civil-Military Relations* (Cambridge, MA: Harvard University Press, 1959).

49. For example, see Shannon E. French, *The Code of the Warrior: Exploring Warrior Values Past and Present* (New York: Rowman & Littlefield, 2003).

50. Huntington, *The Soldier and the State*; Morris Janowitz, *The Professional Soldier: A Social and Political Portrait* (New York: The Free Press, 1960).

Military professionalization presupposes a sharp separation of soldiers from the civilian population. Manuel Davenport correctly observes that "the military profession is bound by a sense of fellowship that is almost inherently anti-civilian."[51] Huntington likewise says that the military maintains a "sense of organic unity and consciousness of themselves as a group apart from laymen."[52] As two sides in a binary, the concepts of soldier and civilian are mutually constituted and therefore mutually threatened. The transformation of the military profession is therefore not restricted to the military. It entails greater difficulty when distinguishing civilians from combatants, both in theory and in practice. When all these changes are put together, changes in the military profession constitute a blurring of the sharp institutional and normative boundaries that help to define conventional interstate war.

Theme 7: Erosion of Norms

Collectively, the foregoing themes raise puzzles about how the use of force can and should be governed, which build a seventh theme: disruption of the prevailing norms of war. During the past century, just war theory has had unprecedented success in being adopted as a normative guide.[53] The principles of discrimination, proportionality, and just cause are especially prominent in international law. According to the principle of discrimination/distinction, it is not permissible to attack civilians, their property, or their cultural artifacts (though harming civilians inadvertently as "collateral damage" is treated as being excusable). Proportionality, which forbids belligerents from using excessive force against their enemies, is a vaguer guideline, yet it is influential. It is now generally accepted that belligerents may not attack retreating or defeated enemies and that they cannot employ many of the immensely powerful weapons that dominated the world wars: massive artillery bombardments, carpet bombing, chemical weapons, and nuclear weapons. The principle of just cause was affirmed by the Nuremberg and Tokyo tribunals, which proscribed the crime of aggression, and thereafter by the United Nations charter.[54] Wars are only permissible when

---

51. Manuel M. Davenport, "Professionals or Hired Guns? Loyalties Are the Difference" In *Military Ethics: Reflections on Principles*, ed. Malham M. Wakin (Washington, DC: National Defense University Press, 1987), 10.

52. Huntington, *The Soldier and the State*, 10.

53. Michael Walzer, *Arguing About War* (New Haven, CT: Yale University Press, 2004).

54. Kirsten Sellars, *'Crimes Against Peace' and International Law* (Cambridge: Cambridge University Press, 2013).

they are waged in self-defense or defense of others. Even though aggressive wars persist, they must do so in the guise of defensive operations, which at least tacitly acknowledges the morality of war. States that fight without trying to demonstrate compliance with just war precepts may be punished by condemnations, economic sanctions, or diplomatic isolation.

The technological, organizational, and tactical/strategic changes I discuss have been shaped by the just war tradition and enthusiasm for using its principles to govern wars. The case studies show that recent innovations have been largely directed at developing modes of fighting that can be effective within the confines of just war theory and international law.[55] Just war theory also provides the language that is used to evaluate, and often to justify, innovations. There has been moral improvement in terms of lowering the overall casualties inflicted and limiting when wars can be waged. Nevertheless, progress is never total because of the tendency for any attempt to restrict war to create new difficulties, much as risk-management strategies may improve security in some ways while still producing new risks that test security in unexpected ways.

The normative confusions of twenty-first-century war involve difficulties applying norms of war and a more fundamental challenge of determining when there is a condition of war that would call for the application of those norms. In terms of the former, there are struggles to figure out where new weapons should fit within standards of *jus in bello* that took shape centuries ago and to decide whether reducing the lethality of conflict alters the *jus ad bellum* calculations relating to when wars may be justifiably initiated. In terms of the latter, there is a profound demarcation challenge. Because the line between war and peace is ambiguous, we experience varying degrees of hostility and the proliferation of war in alternate forms, such as counterinsurgency, peacekeeping operations, hybrid wars, and gray zone conflicts. These types of violence are sustained by uses of military force short of large-scale conventional military actions and often involve work that is more typical of policing. Enemies in these conflicts are likewise treated as falling somewhere between the categories of combatant and criminal. States tend to slide easily between these classifications, even in the midst of wars, using the concepts and their distinctive normative commitments as weapons in their own right. Following from this, there is confusion between normative frame-

---

55. David A. Koplow, *Nonlethal Weapons: The Law and Policy of Revolutionary Technologies for the Military and Law Enforcement* (Cambridge: Cambridge University Press, 2006); Koplow, *Death by Moderation*.

works developed for war and law enforcement, which has far-reaching implications for conceptions of politics that depend on a separation of international and domestic affairs.

The Geographical Scope of This Study

My analysis focuses on how Western states wage war. This geographical choice was made primarily based on which countries are the most active in setting the precedent for innovations by creating them and using them. The United States is the foremost example throughout the book, as it is a frequent belligerent, a leader of military innovation, and exerts far more global influence than other armed forces. With a military budget greater than the next ten countries combined and with a long-standing propensity for seeking advantages with technological advancements, it tends to produce and experiment with military innovations more than other countries.[56] Next in order of precedence are the United Kingdom, France, Germany, Israel, and Russia, which join the US in making up the primary cases because they are likewise leaders of innovation and routinely participate in conflicts that establish the standards of what twenty-first-century warfare is in practice. Armed drones, for example, are now widespread, but their introduction and design has been more heavily shaped by the United States and Israel than by the other countries currently using them.

I draw examples from a range of other countries in addition to the primary cases, including some non-Western countries, when they have an important role in shaping a particular domain of innovation. However, I do not focus on these to the same extent as the primary cases for two reasons. First, I want to show how military innovations are a response to, and cause of, an ongoing crisis for the Westphalian system and conventional war. Because that system and its norms were initially built in a Western context, the disruptions are primarily evident there. As I show throughout the case studies, non-Western states are in some cases better suited to adopting innovations without disruption because they lack the same inherited theoretical framework that is often a barrier in the West. This is evident, for instance, from how China has been able to come to terms with cyberwarfare more effectively than some Western countries. Second, the leading non-Western armed forces currently face different challenges than their Western counterparts. For instance, most do not contend with the demands of counterinsur-

56. https://www.globalfirepower.com/defense-spending-budget.asp.

gency operations and nation building, which many Western states have inherited as part of their colonialist past.

Much more could also be said about how violent nonstate actors (VNSAs), such as insurgents and terrorists, have adjusted to technological and organizational changes, as well as how they have instigated these changes. I address some of these points tangentially, while focusing on what these mean for Western states. It is important to restrict the scope of analysis in this way because states face much different demands than VNSAs. For example, some new weapons technologies, such as drones, may be appealing to states and VNSAs alike, though the reasons for their attractiveness differ. In other instances, such as with nonlethal weapons, states may pursue a new technology without VNSAs showing comparable interest. The same can be said for the organizational and tactical/strategic shifts this book covers. They are primarily initiated by states, although states are often responding to challenges presented by VNSAs.

Conclusion

My analysis of twenty-first-century warfare is deliberately synthetic and open-ended. The approach is synthetic in the sense that my contribution comes less from the analysis of any single topic than from linking seemingly disparate innovations together and showing that a common set of themes run through each. This approach makes it possible to get a sense of the interaction between various innovations and to see the effects of these innovations across multiple domains. Tracing the interconnectedness between the topics adds greater depth to research on each, while also highlighting their political implications more effectively than is possible when only focusing on one specific topic. The analysis is open-ended because it considers how fluid themes come together in a loose assemblage without ever producing a definitive model of twenty-first-century warfare. I want to avoid the pitfalls that inevitably follow from attempts to build comprehensive, grand theories of warfare that rely on rigid categories. I reject the view that wars fall into particular types that can be easily categorized, and instead I argue that conflicts of a given era may vary considerably while still evincing themes that establish family resemblances.

Part I • Technological Innovations

1 • War by Proxy

Drones and the Dehumanization of War

The history of warfare reveals a perennial aspiration to strike others while minimizing the chance of being attacked in turn. From bows and slings to bombs and missiles, we can see this goal being pursued by adding *distance* between attackers and their targets. Ranged attacks transform space into a defensive instrument that insulates attackers from reciprocal threats. Even societies that celebrated a warrior ethos rooted in hand-to-hand combat often sought to capture some advantage in distancing fighters from their opponents. The Greek hoplites and Macedonian phalanxes put space between them and their enemies with long dory and sarissa spears; the Roman legions often threw an opening volley of pila before closing for hand-to-hand combat, and knights relied on horses to hold them aloft. However, until the closing decades of the twentieth century, the mediating power of distance was limited by the high costs of fighting at longer ranges (usually in terms of lost accuracy, but also financially and in lost opportunities to develop other military capacities) and by the enemy's ability to match or counter innovations. For example, during the Second World War, bombers became less accurate when trying to protect themselves by flying higher and were vulnerable to comparatively low-cost countermeasures.[1]

For the first time in history, the dream of fighting effectively from a distance and denying adversaries a chance to strike back is within reach. Drones are revolutionary weapons platforms not only when judged against the weapons of conventional war but when judged against weapons of any era, because they allow their users to fight at extraordinarily long ranges while also increasing capacities for selectively engaging specific targets. It is now

1. A. C. Grayling, *Among the Dead Cities: The History and Moral Legacy of the WWII Bombing of Civilians in Germany and Japan* (New York: Walker & Company, 2006).

possible to imagine "risk-free wars" in which fighting is so radically asymmetric that the casualties may be sustained by one side only, and yet without attackers suffering from the accuracy or opportunity costs that have historically been associated with ranged weapons.[2] Drones increase distance *and* accuracy, improve the user's security *and* decrease the opponent's, lower the costs of fighting for allied forces *and* raise them for adversaries. Direct countermeasures do exist. Drones can be shot down or hacked. They may also suffer from technical faults and user mistakes. Current-generation drones are arguably more susceptible to these problems than manned vehicles, as they tend to be fairly primitive by comparison.[3] Nevertheless, unlike weapons and vehicles of previous generations, the fate of the drone is completely detached from that of the user. Any attack against a drone, even one that succeeds in destroying it, leaves the user unscathed. For this reason, drones can simultaneously maximize force protection and offensive power.

In this chapter, I discuss drones as a general class of weapons platforms. This class is epitomized by the unmanned aerial vehicles (UAVs) that have achieved prominence in American targeted killing operations, such as the MQ-1 Predator and MQ-9 Reaper, but it is important to look beyond UAVs to see the scope of the change that drones will bring. A variety of machines operating in the air, on land, and at sea are poised to expand the scope of drone operations in coming years. These weapons platforms will vary a great deal, as they must to take on a broad range of different roles, but they share the capacity for acting as proxies that sever the physical connection between machine and user. That is to say, drones are distinguishable as a class of machines that hold the promise of achieving perfect force protection through distanciation while also improving effectiveness in combat. This does not necessarily assume that drone operators will always be thousands of miles away from the battlefield. Tactical drones and bomb disposal drones may only be operated from short distances. What is important is not the magnitude of the distance but rather that the machines grant the user greater security by breaking the physical link between the platform and its user in such a way that the latter can choose to remain just beyond the enemy's reach.

Commentators on both sides of the debate over whether drones are justi-

2. I use the terms "symmetric" and "asymmetric" to refer to the balance of power between opposing belligerents. Wars are symmetric when the opposing sides fight in roughly the same ways and are close enough in relative power that both sides have a reasonable hope of victory when fighting according to the same rules. Wars are asymmetric when the opposing sides fight in different ways or when there is a vast disparity in military strength.

3. Michael C. Horowitz, Sarah E. Kreps, and Matthew Fuhrmann, "Separating Fact from Fiction in the Debate over Drone Proliferation," *International Security* 41, no. 2 (2016): 7–42.

fiable weapons have offered competing labels for these machines. Air forces have attempted to rebrand them as "remotely piloted aircraft systems (RPASs)," "unmanned aircraft systems (UASs)," or "remotely piloted aerial vehicles (RPAVs)," while opponents propagate more pejorative names like "killer drones" or "assassination drones." In each case, these names are clearly motivated by efforts to convey a certain impression of the machines. There are ostensibly more neutral terms, such as unmanned aerial vehicle (UAV) or tele-operated aircraft, but these tend to come with misleading connotations. For example, talking about UAVs obscures the deep affinities these have with machines based on land or at sea. I prefer the term "drone" for three reasons. First, the fact that it is distasteful to many commentators on both sides of the debate suggests that it is fairer than the more overtly charged labels advanced by one side or the other. Second, it does not distract from the family resemblances unifying drones of various types by focusing narrowly on a certain subset of these machines, such as those operating in the air. By my account, the various types of drones are more alike than they are different because they share the quality of acting as proxies for human combatants. I therefore reserve the term "UAV" for times when I am dealing specifically with the aerial drones that have so far provided most of the empirical evidence on how drones operate, and otherwise refer to drones as a general category operating in any terrain. Finally, "drone" is preferable because it is the most familiar term and the one that has had the most widespread usage across academia, government, the media, and the general public.

The Rise of Drones

The history of drone warfare can be traced to the Second World War. Remote-controlled ground vehicles like Germany's Goliath tracked mine were created to destroy armored vehicles from a distance, while American Radioplane aerial drones helped to train pilots in combat maneuvers.[4] Ground-based drones were unsuccessful, as they were unreliable, had to be controlled at close range, and were relatively expensive. Interest in drones persisted throughout the twentieth century, but the major innovations came in the aerial domain. The American Firebee gradually transitioned from a target practice tool akin to the Radioplane to carrying surveillance equipment over enemy territory. During the Vietnam War, remote-

4. Paul Fahlstrom and Thomas Gleason, *Introduction to UAV Systems* (Hoboken, NJ: Wiley, 2012).

controlled aircraft flew 3,435 reconnaissance sorties and became more advanced with the addition of countermeasures to deflect incoming missiles.[5] These drones never reached the level of technical sophistication needed to be reliable weapons platforms, as they had to be controlled by DC-130 director aircraft flying in the same airspace and could not operate independently if their link to the operator was severed.[6] It was not until the 1980s and 1990s that advances in telecommunications, global positioning systems, and artificial intelligence made drones viable instruments of war.

The United States Air Force developed the RQ-1 Predator as a long-range surveillance machine, using it to locate enemy positions and conduct bomb damage assessments during the Gulf War and the intervention in Kosovo. In 2000 it began flying them over Afghanistan in the hunt for Osama bin Laden following his attacks on US embassies in Kenya and Tanzania in 1998. The goal was to use drones to identify the terrorist leader for attack by a cruise missile. It was at this critical juncture, with unarmed drones and cruise missiles paired in joint operations, that the limitations of both became evident. Cruise missiles could take minutes to reach a target, long after the intended victim may have left. Drones could provide real-time intelligence, but without the ability to launch attacks, the value of this information was limited. A breakthrough came in 2001 when the Central Intelligence Agency took the step of arming the aircraft with two Hellfire missiles and using them to launch attacks against suspected terrorists.[7] The first attack came against Mullah Omar, the leader of the Taliban. Although it only succeeded in killing a few bodyguards, this proof of concept built support for subsequent missions. It also illustrated the overall arc of drone innovation. Machines that were tentatively introduced as one of the many options for increasing lethality during Westphalian conventional wars evolved into machines for the kind of extrajudicial violence against terrorists that pushes war beyond conventional boundaries.

The introduction of the MQ-9 Reaper in 2007 marked another turning point, as the machine was designed to carry much larger weapons payloads and to launch multiple strikes. It was evidence of a policy decision to pursue drone warfare with a machine capable of working in "a true hunter-killer

5. Fahlstrom and Gleason, *Introduction to UAV Systems*, 5.

6. Thomas P. Ehrhard, *Air Force UAVs: The Secret History* (Arlington, VA: Mitchell Institute for Airpower Studies, 2010).

7. Chris Woods, "The Story of America's First Drone Strike," *The Atlantic*, May 30, 2015, http://www.theatlantic.com/international/archive/2015/05/america-first-drone-strike-afghanistan/394463/

role."[8] Under the Obama administration, drones became one of the defining counterterrorism instruments. As Woods explains, "Back in October 2001, the United States had been able to field just one armed Predator above Kandahar. By 2014, it could fly 63 simultaneous combat air patrols 24 hours a day in the skies of half a dozen countries."[9] The Predator and the more advanced Reaper were used in hundreds of targeted killings in Iraq, Afghanistan, Yemen, Pakistan, Somalia, Libya, and the Philippines as well as ground support operations in Afghanistan and Iraq. These missions demonstrated drones' versatility and set an international precedent for their use.

The United States receives a disproportionate share of the attention for conducting drone operations, but it is not alone. Israel has deployed drones in similar roles. Although its operations have been limited to the Palestinian territories and neighboring countries, they have persisted longer. Israel used UAVs to provide intelligence and designate targets during fighting in the Bekaah Valley in 1982,[10] and it continues to use drones in surveillance and targeted killing operations in the West Bank and Gaza.[11] Moreover, unlike many other drone users that depend on American support (such as the countries of western Europe), Israel produces its own machines, including the IAI Heron and IAI Eitan. It sells these internationally, making the technology more open to militaries around the world and limiting the United States' ability to exercise control over the market as diplomatic leverage.[12]

The United Kingdom has acquired its own Reapers for long-range targeted strikes and sometimes works alongside American crews, as in the attack on Jihadi John in 2015.[13] It reported flying more than 100,000 hours in Afghanistan as of 2012, firing 349 missiles.[14] By the end of September 2020,

---

8. "'Reaper' Moniker Given to MQ-9 Unmanned Aerial Vehicle," Air Force, 2006, https://www.af.mil/News/Article-Display/Article/129780/reaper-moniker-given-to-mq-9-unmanned-aerial-vehicle/

9. Chris Woods, *Sudden Justice: The True Costs of America's Secret Drone War* (London: C. Hurst & Company, 2014), 179.

10. Matthew M. Hurley "The Bekaa Valley Air Battle," *Airpower Journal* (Winter 1989).

11. Yaakov Katz and Amir Bohbot, *The Weapon Wizards: How Israel Became a High-Tech Military Superpower* (New York: St. Martin's Press, 2017); Ronen Bergman, *Rise and Kill First: The Secret History of Israel's Targeted Assassinations* (New York: Random House, 2018).

12. http://www.jpost.com/Israel-News/India-to-receive-armed-Heron-drones-from-Israel-486016

13. Gordon Rayner, "How the US and UK Tracked Down and Killed Jihadi John," *The Telegraph*, November 13, 2015, https://www.telegraph.co.uk/news/worldnews/islamic-state/11993569/How-the-US-and-UK-tracked-down-and-killed-Jihadi-John.html

14. http://www.publications.parliament.uk/pa/cm201213/cmhansrd/cm121106/halltext/121106h0001.htm#12110684000252

it had conducted more than four thousand Reaper missions in Iraq and Syria.[15] Italy has likewise been eager to take advantage of UAVs and has conducted joint training with American pilots,[16] However, it has been hindered by US unwillingness to arm Italian drones and a lack of domestic drone-building capacities. Despite years of debate and considerable domestic opposition, Germany began leasing Heron TP drones from Israel in 2018.[17] This is a sign that even countries skeptical about the future of drone warfare are apt to be carried along with the trend as they strive to maintain modern armed forces. France has a mixed fleet of Reaper and Harfang drones. It launched its first drone strike against terrorists in Mali in 2019.[18] Drone activity across the European Union is likely to change in the near future following joint agreements to invest in the production of drones in Europe.[19] Although it initially lagged behind the United States and Israel, Russia has emerged as one of the most enthusiastic drone users. It has conducted strikes in Syria and Libya, and it is exploring ambitious plans for the future, such as heavy bombers and tanks that can be controlled remotely or by autonomous systems.

The ongoing Libyan civil war has become the most active drone battlefield and a prime example of the complexity of post-Westphalian conflict. At least five different countries (the United States, Turkey, the United Arab Emirates, Egypt, and France), as well as domestic actors, have launched hundreds of attacks per year using UAVs.[20] Those on both sides claim to have shot down rival UAVs, and while it is not clear whether these claims are accurate, photographers documented several downed drones per month throughout 2020. That it is unclear who is responsible for particular attacks, whether certain countries are involved, and whether the drones have been shot down or crashed shows how effectively participants are able to conceal their mediated violence.

Drones are even attractive to weak states and nonstate actors. Hezbollah has used remote-control aircraft to conduct reconnaissance over Israel. Mexican drug cartels have used them to transport drugs across the border. The

---

15. https://dronewars.net/uk-drone-strike-list-2/

16. Tom Kington, "Italian Reaper Drones To Be Used for Crowd Monitoring" *Defense News*, December 17, 2014, http://www.defensenews.com/story/defense/international/europe/2014/12/17/italian-reaper-drones-to-be-used-for-crowd-monitoring/20528495/

17. https://www.efadrones.org/

18. https://www.defense.gouv.fr/operations/barkhane/breves/barkhane-poursuite-de-l-action-dans-la-region-de-mopti

19. Andrew Rettman, "Seven EU States Create Military Drone 'Club,'" *EUObserver*, November 20, 2013, https://euobserver.com/defence/122167

20. Samuel Oakford, "Libya's Lawless Skies," *Airwars*, June 20, 2018.

Islamic State made widespread use of small drones to carry light bombs as well as to direct mortars.[21] The proliferation of relatively inexpensive hobbyist UAVs is evidence that the technology will continue to be accessible, potentially providing a new route for delivering improvised explosive devices. In short, interest in drones is growing and an arms race is underway, signaling that drones will be a fixture of conflicts in the coming decades. Although many drones are envisioned as having conventional war roles, the missions they have performed thus far shows that much of the enthusiasm for these weapons platforms lies in their utility as coercive instruments across a broad spectrum of conflict scenarios.

Riskless Warfare via Mediation

Drones mark a radical departure from the past, yet they are informed by a perennial strategic interest militaries have in force protection: the minimization of danger to friendly personnel. Force protection is vital, especially when victory is determined by a belligerent's ability to outlast the enemy in pitched battles. Militaries seek force protection in various ways, such as with superior training, speed, maneuverability, and strategy. The history of technological and organizational innovation in war is to a large extent a story of pursuing force protection. Mediation is among the most common and effective approaches to insulating soldiers from danger, and it is generally facilitated by spatial distance and armor.

Throughout the twentieth century, a rapid proliferation of indirect fire weapons was guided by the effort to limit threats with the help of distance and increased armor. War went through a rapid spatial expansion, first with long-range artillery, then aerial bombardment, then missiles. By the end of the century, it was possible to fight over unlimited geographical expanses. Armor did not develop at quite the same pace, yet improvements to it were nevertheless profound. Throughout the first half of the twentieth century, belligerents sought to produce larger and more heavily armored ground vehicles and ships, then, reaching the limits of size, they spent the following decades developing new composite materials that could improve protection without sacrificing speed and operating range.[22]

---

21. Anthony Capaccio, "Extensive Islamic State Drone Use Raising Risks in Mosul Battle," *Bloomberg*, October 26, 2016, https://www.bloomberg.com/news/articles/2016-10-26/extensive-islamic-state-drone-use-raising-risks-in-mosul-battle

22. Paul J. Hazell, *Armour: Materials, Theory, and Design* (Boca Raton, FL: CRC Press, 2016).

Until the twenty-first century, improvements in force protection invariably came up against a major barrier: any advantages achieved via physical mediation tended to be short-lived because these were counteracted by new threats. Innovations were matched by opponents, prompting still more efforts to improve ranged fighting capacities or armor. The result was continual arms races as new weapons systems were introduced, culminating in the apocalyptic quest to build massive arsenals of nuclear weapons. Those who could not compete directly by building their own variants of the latest machines took the alternative path of investing in cheaper countermeasures that exploited technical limitations.[23] Although its own forces were inferior to those of the United Kingdom, Argentina was able to strike British ships with its Exocet missiles during the Falklands War. The US Navy is at risk of a similar fate from Chinese Silkworm missiles deployed to protect the South China Sea. Israeli tanks easily triumphed over the Soviet machines used by its Arab neighbors during the Six Day War in 1967, only to find themselves struggling for survival when Egypt deployed Sagger antitank missiles during the Yom Kippur War in 1973.[24] Even violent nonstate actors can counter the most advanced force protection methods using distance and armor. During the Battle of Mogadishu in 1993, the US military found that mediation through distance via Black Hawk helicopters could be countered by rocket-propelled grenades armed with timed fuses, and that mediation through the armor of Kevlar vests could not ensure complete protection against small arms fire at close ranges.[25] The lesson was that advances in force protection could be matched by rivals and counteracted by less powerful adversaries. This is the familiar pattern of the conventional war arms race.

The invulnerability of UAVs may seem to be a characteristic of unconventional warfare, rather than of the machines themselves, because UAVs are currently used in contexts in which manned aircraft would likewise be nearly invulnerable. In his defense of the morality of drone warfare, Bradley Strawser argues that the risk asymmetry between a manned F-22 and a tribal warrior is roughly the same as that between a drone and a tribal warrior.[26] The presence of a human in the cockpit makes little difference when the tribal

23. Edward N. Luttwak, *Strategy: The Logic of War and Peace* (Cambridge, MA: Belknap Press, 2001).

24. Abraham Rabinovich, *The Yom Kippur War* (New York: Schocken, 2004).

25. This was partially because many American soldiers did not equip themselves with all the body armor available. They opted to increase their mobility and lighten their equipment load.

26. Bradley Jay Strawser, "Moral Predators: The Duty to Employ Uninhabited Aerial Vehicles," *Journal of Military Ethics* 9, no. 4 (2010): 342–68.

warrior is unable to shoot down aircraft of any kind. The two aircraft are, in this scenario, virtually identical when it comes to the risk to pilots. This is borne out by how aerial campaigns are conducted in the absence of armed drones. The intervention in Kosovo in 1998 was carried out exclusively by air to avoid putting ground forces at risk. That campaign bears many similarities to drone strikes throughout the War on Terror, with Western forces being able to attack at will without endangering soldiers. With this in mind, it would be misleading to suggest that radical risk asymmetries are unique to drones. Extreme risk asymmetries have existed in conflicts since the end of the Cold War, and they would persist even if drones were removed from service because the world's leading military forces have achieved substantial advantages across a spectrum of offensive technologies.

What really sets drones apart from preceding military technologies is not the goal of force protection or even how this goal is pursued through mediation. Rather, the novelty comes from the aspiration to assert total control over risks, such that drone operators are safe from any conceivable response. Unlike other weapons that may establish brief asymmetries between technologically mismatched opponents, drones act as human proxies in every setting. They sever the link between operator and machine. They allow their users to conduct surveillance, move supplies, fight, and perform any number of other tasks without exposing the operators to enemy fire. By disassociating operators and machines, even the most hazardous missions can be performed without any personnel losses. Strawser is certainly right in thinking that an F-22 and a UAV may be equally invulnerable when attacking Taliban insurgents in the mountains of Afghanistan, yet this analogy depends heavily on the assumption that enemies always lack the ability to shoot down aircraft. This is certainly not the case. Some armed forces do have the capacity to kill the pilot of an F-22 but would be unable to kill a UAV pilot by attacking the UAV itself. Whereas any increase in risk to the F-22 is necessarily an increase in risk to the pilot, increasing the risk to a drone does not affect the risk to the operator. Drone operators remain safe regardless of how effectively enemies are able to destroy the machines. Thus, a distinctive new type of mediation characterized by machines acting as proxies for human combatants brings twenty-first-century warfare into a post-Westphalian domain of risk management in which entirely risk-free attacks are inherent in weapons platforms, rather than simply being a feature of ephemeral asymmetries between the opposing belligerents.

Just as with previous military innovations, drones will doubtless provoke arms races and the creation of countermeasures. They will not always grant their users a permanent advantage any more than previous innovations have.

UAVs and other drones are apt to be destroyed in much higher numbers in conventional conflicts than when fighting against insurgents. However, regardless of whether countermeasures are directed against drones, the fate of the operators will remain divorced from that of the machines. Physical separation from the machines ensures that operators will be safe even when their drones are attacked; any attack aimed at killing the operators will have to be directed against them specifically, and not against the weapons platforms they control. With drone operators usually being positioned far from the battlefield, and the likelihood of drones gaining autonomy to the extent that humans are less directly involved in their operations, belligerents will find it difficult to attack enemy drone operators. This means that in any kind of conflict, even one between conventional adversaries, drone users will never face the same prospect of sustaining casualties that they once did. Any damage to the drones will be measurable in terms of money and machines alone. This is a decisive triumph of force protection via physical mediation, which gives drones a much different character than any preceding weapon.

Existential and Instrumental Warfare

War takes on profound existential meaning when it is treated as an end in itself or as an activity with inherent value that is passed along to the participants. The identity of the military profession is bound up with an existential notion of war as a heroic enterprise, as are many societal normative frameworks associated with nationalism and patriotism. However, from a political standpoint, war is largely instrumental; it is a means of securing political objectives, rather than an activity with self-contained meaning. This is perfectly captured by Clausewitz' dictum that "war is not merely an act of policy but a true political instrument, a continuation of political intercourse, carried on with other means,"[27] which is the purest exhortation for conflict to serve higher ends beyond itself. Instrumentalism is usually the optimal choice for policymakers because war is politically risky. Resorting to violence can ruin states, or at the very least ruin the careers of government officials. The more a polity decries the evils of war and claims to pursue violence only for self-defense or humanitarian concerns, the more the instrumental logic comes through. From this perspective, war can only be legitimate in light of some higher goals that redeem it.

Contemporary discourse about war is rife with contradictions and para-

27. Carl von Clausewitz, *On War* (Oxford: Oxford University Press, 2007), 28.

doxical thinking when navigating these divergent perspectives on whether war creates its own meaning or only becomes meaningful in light of political goals. This is because of how these contrary perspectives were forced to coexist during the conventional wars of the nineteenth and twentieth centuries, with instrumentalism in senior positions and existentialism in the ranks. The instrumental subordination of war to politics holds many attractions, especially in an era when destructive power has reached a point that it is difficult to imagine governments fully embracing an existential conception of war without this causing rampant, unnecessary violence that results in global devastation. However, instrumentalism presents its own problems. The political goals being pursued are often arcane or inconsequential from the perspective of the individual soldiers who do the fighting and bear much of the risk. Many soldiers show little interest in the political considerations at hand and instead focus on protecting their comrades and on the experience of fighting as an end in itself.[28] It has become a cliché that soldiers are really "fighting for each other." Indeed, the professional armed forces of contemporary Western democracies actively discourage soldiers from developing a stake in wars as political events. Soldiers are supposed to be unbiased servants of the state, in essence, instruments in the state's instrumental use of violence to achieve political objectives, who are themselves beyond instrumental thinking because they are invested in the depoliticized experience of fighting. The more war drifts toward the instrumental ideal, the more its romanticism is stripped away and the more difficult it becomes for participants and civilian audiences to derive a deeper sense of meaning or identity from the fighting.

Armed forces continue to endorse an existential conception of war as a cornerstone of the profession and its norms. Doing so is advantageous when it comes to maintaining civilian political control, as it helps ensure that the military will to wars because of their inherent duty to do so, rather than questioning the goals being pursued. Moreover, existentialism is vital for giving those who bear the most personal risk a sense that they are achieving something that truly warrants the loss of their lives. Military professionals see themselves as warriors, and warriors do not shy away from conflict or death. Continually facing one's own mortality is celebrated by those who see war as a test of virtue or an ennobling experience. As Christopher Coker correctly notes, "The reason that warriors have fought war so enthusiastically is that it

28. Charles C. Moskos Jr., *The American Enlisted Man: The Rank and File in Today's Military* (New York: Russell Sage Foundation, 1970), 146.

has offered them an existential experience found in no other walk of life."[29] Soldiers are often described as exemplars of virtue and patriotism precisely because they routinely face death. "The myth of war entices us with the allure of heroism."[30] This sensibility then extends beyond the profession as its members are used as a reference point for all patriotic citizens.

The methods of managing risks have hitherto left the existential conception of war largely unchallenged, allowing it to coexist with the instrumental view throughout the Westphalian era. Governments were able to pursue wars for political objectives; armed forces could take a depoliticized pride in the glory of patriotic violence. Technological innovations that reduced risk through physical mediation coexisted with an evolving sense of the warrior ethos because risk was never totally averted. Some of the most advanced weapons even produced their own compelling sense of danger. We can see this from fighter pilots, who, though they are often akin to drone operators in their insulation from enemy attacks, are romanticized because of the excitement of aerial combat and the hazards of piloting aircraft. Attempts to minimize danger may contribute to war's appeal, so long as the danger is not completely removed. Soldiers' narratives are replete with vivid expectations of what it will be like to fight and kill, which are often reminiscent of teenagers' anticipations of a first sexual experience. Efforts to reduce risk have made war a safer fantasy to explore, and one that is all the more mysterious because of efforts to force it into a hermetically sealed space divorced from civilian life.[31]

War must be risky to preserve its existential character. Warrior excellence can only be achieved by enduring danger and defeating enemies in an evenly matched test of skill. However, there also needs to be some chance of surviving combat. When the horrors of war threaten to overwhelm soldiers, virtually ensuring death or horrific injury, combat loses its mystique. Soldiers may mutiny, as exhausted French and Russian forces did in 1917, or they may surrender without putting up the fight their political masters' demand, as dozens of German and Russian units did when they were isolated during the bitter fighting of the Second World War. Militaries and the governments they serve therefore strive to strike a balance when promoting war as an edifying form of self-sacrifice in pursuit of higher ideals. Danger must exist, but it cannot be too great. Subtle interventions to manage risk before the twenti-

29. Christopher Coker, *Waging War Without Warriors?* (Boulder: Lynne Rienner, 2002), 30.

30. Chris Hedges, *War Is a Force That Gives Us Meaning*, 83.

31. Anthony Swofford, *Jarhead: A Marine's Chronicle of the Gulf War and Other Battles* (New York: Scribner, 2003).

eth century were always incomplete; they mitigated danger to soldiers without eliminating it. This balance was not by design. It was a default position because belligerents lacked the means to eliminate the risk of being attacked by enemies who were themselves caught up in the same balancing act.

The symbiosis between existential and instrumental conceptions of war began to break down in the mid-twentieth century, when soldiers' lives became too valuable to be wantonly risked. The shift to all-volunteer recruitment cut armed forces off from an assured supply of conscripts to fill the ranks. Ironically, advancements in mediation aimed at protecting soldiers also helped to make soldiers more valuable by increasing the costs of training them and replacing those who were lost in combat. In the twenty-first century, soldiers are not merely bodies to throw at the enemy in waves, but are skilled technicians who represent a large expenditure of resources. This has directly affected the overall appetite for risk in armed conflicts. As Hugh Smith points out, "In broad terms the more difficult it is to raise and maintain armies, the more reluctant are military (and political) leaders to risk large numbers of lives."[32] Soldiers may still seek the existential experience of war—the danger of facing enemies on the battlefield and risking death—but that experience is becoming far too costly from a political standpoint. Each combat death constitutes a massive lost investment and poses the challenge of finding recruits who will be willing and able to fill the vacancy. Moreover, the dwindling size of armed forces means that even slight casualties can have a jarring impact on a military's effectiveness.

Although military personnel continue to define themselves according to a "warrior ethos" and express a desire for the decisive combat of conventional operations, the costs of indulging this desire have become exorbitant, often too high to justify the risk from an instrumental perspective. The costs have risen not because of an external challenge from nonstate actors but because of actions taken by states themselves. Their own pursuit of more advanced conventional weapons raised the price of training and the destructiveness of fighting to the point that conventional war can no longer be treated as a standard form of interstate dispute resolution, as it was during the Westphalian era. Drones have become one of the most compelling options for protecting the investment in military expertise. Rather than moderating risk exposure, they completely eliminate it. They are attenuating the existential conception of war by physically removing military personnel from any situations in which they might be at risk of injury or death. To the extent that drones

32. Hugh Smith, "What Costs Will Democracies Bear? A Review of Popular Theories of Casualty Aversion," *Armed Forces & Society* 31, no. 4 (2005): 487–512, 491.

spread into combatant roles associated with existential violence, they undermine a core component of the military profession and its norms.[33]

Changing Perceptions of War

War can be a paradoxically unifying experience for opposing combatants because of their shared experience of the existential dimensions of violence.[34] Past wars offer many instances of soldiers temporarily putting their differences aside in displays of respect, such as by refusing to shoot at helpless adversaries, trading food, and establishing informal truces.[35] Even fairly recent conflicts like the Vietnam War and the Falklands War produced some striking moments of mutual recognition between adversaries.[36] This feeling does not always emerge; there are certainly cases in which no respect is evident because the enemy's humanity is thoroughly denied. However, the possibility of recognition between opponents remains a factor that at least potentially limits the ferocity of violence and paves the way for postwar reconciliation. In this sense, the existential conception of war sustained by military culture can play an invaluable role. It provides a basis for mutual recognition between combatants that parallels a politics of mutual recognition, even in the midst of hostilities, which the Westphalian system depends on.

The relationship of recognition is undermined by the changes to the military profession instigated by each of the transformations I discuss in this book, but it is particularly clear with respect to drones. By separating weapons and their operators, drones cause a phenomenological shift that redefines the experience of fighting as inherently asymmetric. Drones are "overcoming the unity of killing and dying embodied in the classical type of soldier,"[37] thereby creating a relationship that is instrumental, rather than existential. The UAVs currently in service are largely invisible to the people they attack. Drone pilots watch their targets silently and send missiles without warning.

33. The challenge that drones present to norms of war and the military profession is primarily associated with armed drones, as they are the machines supplanting warrior roles in combat. Unarmed drones are typically used in roles that lack strong links to the warrior ideal and are therefore less threatening to it.

34. Joanna Bourke, *An Intimate History of Killing: Face-to-Face Killing in Twentieth-Century Warfare* (London: Granta Books, 1999).

35. Tony Ashworth, *Trench Warfare 1914–1918: The Live and Let Live System* (London: Macmillan, 1980).

36. Ted van Baarda, "The Ethical Challenges of a Complex Security Environment," in *Ethics, Law and Military Operations*, ed. David Whetham (London: Palgrave, 2011), 148–72.

37. Herfried Münkler, *The New Wars* (Malden, MA: Polity Press, 2005), 132.

They project military power into every space that is visible from the air, giving enemy fighters reason to fear attack at any moment. Gregory argues that "these new visibilities produce a special kind of intimacy that consistently privileges the view of the hunter-killer, and whose implications are far more deadly."[38] Pushing this point further, we should recognize that this perspective is guided not by the same warrior mentality that one might find in previous wars but by a different sort of orientation that has repudiated killing as an edifying contest of martial skill and reconceptualized it as a means to an end. The drone operator's orientation is to kill enemies as efficiently as possible, often at the behest of targeting criteria set out by political leaders and directly supervised by intelligence operatives.

The methods of employing drones will likely change in upcoming conflicts, and it is entirely possible that adversaries will face off with comparable drone forces at their disposal. Nevertheless, even if there is a symmetry of capacities, such as in a war between two states with drone fleets, the perceptual asymmetry will remain the same because it is built into the function of drones as human proxies. If both sides were to possess drones, combatants would not face each other directly but rather would participate in opposing efforts to inflict mediated violence. Attackers will see their contribution to the war effort through the mediation of a machine that stands between them and their human opponents, absorbing the risks of fighting. For those being attacked, the experience of war will be marked less by the intense terror of coming under enemy fire and more by the paranoia of being under constant threat. Perhaps the best chance of some shared experience would come from drone operators on opposing sides being simultaneously hunters and hunted, yet in this case common ground would be undermined by their physical distance and the machine mediation that prevents them from having direct contact. Thus the conquest of risk via physical mediation not only disrupts the military profession and its norms but also disrupts the perceptual experience of war in such a way as to create a barrier against extending norms to adversaries as a form of mutual recognition.

Many critics have said that the risk asymmetry drones produce is inherently immoral because it denies the less powerful side its right to self-defense or forces it to deviate from the norms of war in a desperate bid to win. To be clear, I disagree that there is anything morally objectionable about asymmetry. There is nothing in the just war tradition or international law that

38. Derek Gregory, "From a View to a Kill: Drones and Late Modern War," *Theory, Culture, & Society* 28, no. 7–8 (2011): 188–215, 193.

requires war to be fair and symmetrical.[39] Moreover, the right of self-defense does not disappear because of asymmetry. That right grants an authorization to act in self-defense; it does not, and cannot, guarantee that the defender has a fair chance of winning. The extreme asymmetries of drone warfare leave the right of self-defense intact even as drones make the practical challenge of exercising it effectively more onerous. The issue of asymmetry that I am raising here is therefore not a moral concern in itself but rather an issue about how asymmetry is apt to shape the perception of conflict by increasing the prominence of the instrumental conception of war over the existential.

If drones limit the basis for mutual recognition between opposing combatants, then some new understanding of the enemy must emerge to replace it. Critics of drones argue that mediation puts pilots at risk of mistaking war for a video game and losing any sense of the consequences of their actions, in which case those on the receiving end of the violence are not only denied recognition as warriors but even as humans.[40] Furthermore, this loss of "the real" in war could arguably make it easier to kill and lower inhibitions against wanton destruction. Other commentators reach the opposite conclusion, saying that because drones mediate war with high-resolution cameras that provide an intimate look at targets, they actually highlight the realities of war far more than other ranged weapons that are already in use (artillery, bombs, etc.).[41] This is an open debate that cannot be settled until we have far more empirical research on operators' perceptions. Nevertheless, even without a conclusive answer, the debate is revealing. With such intense disagreement over whether combatants still experience war as being real, it is clear that we have reached a new stage in our understanding of what war is as a perceptual experience. Fighting has become so thoroughly mediated that it is conceivable for soldiers to do it without being aware.

The debate over the effects of mediation shows that the quest to eliminate risk to soldiers via mediation has led us to a point that mediation itself is a risk that armed forces must guard against. Matt Martin says that watching combat from a Predator is "Almost like playing the computer game Civilization, in which you direct units and armies in battle. Except with real

---

39. Marcus Schulzke, "The Morality of Remote Warfare: Against the Asymmetry Objection to Remote Weaponry," *Political Studies* 64, no. 1 (2016): 90–105.

40. Caroline Holmqvist, "Undoing War: War Ontologies and the Materiality of Drone Warfare," *Millennium* 41, no. 3 (2013): 535–52.

41. Ronald C. Arkin, *Governing Lethal Behavior in Autonomous Robots* (Boca Raton, FL: Taylor & Francis Group, 2009); Marcus Schulzke, *The Morality of Drone Warfare and the Politics of Regulation* (London: Palgrave, 2017).

consequences."[42] To counteract this tendency, commanders remind drone operators that they are waging real wars and that they are killing real people.[43] They affirm the reality of war in an effort to avert the risk of war becoming too much like a simulation. It seems unlikely that drone operators will actually lose their sense of reality, as some critics fear. However, it is unlikely precisely because armed forces remind drone operators that their heavily mediated actions have real consequences. The introduction of drones has made it necessary for armed forces to guard against the risk of dissociative thinking. This form of risk management is a profound statement about how much drone strikes differ from previous forms of warfare. It shows that war, at least where drones are involved, is being deprived of inherent meaning to such an extent that the participants need to have their perceptions managed by superiors higher up the chain of command. In other words, violence deprived of its existential meaning must be reimbued with meaning from superiors who are thinking in instrumental terms.

The instrumental goals of war should come from citizens and their representatives, but the perceptual changes associated with drones may arguably change how the former participate in security decisions. A number of commentators have claimed that the media image of war is superseding the reality of war in the public consciousness.[44] Essentially these critics worry about drones and other forms of mediated warfare causing civilian audiences to lose sight of the costs of violence. These conclusions are unwarranted based on the available empirical evidence. Drone strikes have not produced political disengagement, as evidenced by voluminous writings from academics and journalists, as well as the extent of antidrone activism.[45] Some research has even shown that drones have a sensitizing effect, making civilian audiences more attuned to the costs of violence.[46] Drones are a contentious, high-profile political issue precisely because mediation does not produce the kind of disengagement that so many critics have predicted. The real problem is not that civilian audiences are indifferent to the costs of drone attacks but that

---

42. Matt J. Martin and Charles W. Sasser, *Predator: The Remote-Control Air War Over Iraq and Afghanistan: A Pilot's Study* (Minneapolis, MN: Zenith Press, 2010), 31.

43. Martin and Sasser, *Predator.*

44. Jean Baudrillard, *The Gulf War Did Not Take Place* (Bloomington: Indiana University Press, 1995); Colin McInnes, *Spectator Sport War: The West and Contemporary Conflict* (Boulder: Lynne Rienner, 2002).

45. Medea Benjamin, *Drone Warfare: Killing by Remote Control* (New York: OR Books, 2012).

46. James Igoe Walsh, "Precision Weapons, Civilian Casualties, and Support for the Use of Force," *Political Psychology* 36, no. 5 (2015): 507–23.

they must rethink how to evaluate those costs. Civilians tolerated high casualties in conventional wars because they were brought into the existential spirit. They were told to venerate military sacrifice as a form of patriotism and to see armed forces as guarantors of freedom. The instrumentalism of drone strikes puts civilians into a much different decision-making calculus, in which violence often appears to be more callous and less meaningful. In particular, it is difficult to see a direct link between the killing and the protection of domestic life.

Christopher Coker has made strong claims about the adverse consequences that follow from war becoming more heavily defined by its instrumental character. While I generally agree with his account of how instrumentalism is increasing, I contend that his interpretation of the political implications of this process are incorrect. Coker argues that "Modern soldiers are the products of a purely visual culture . . . they are creatures of an age that has made violence largely 'virtual' and war an entertainment that raises few questions, moral or political."[47] On the contrary, we should expect war to become more heavily politicized as drones, as well as the other innovations I discuss in later chapters, deprive war of its existential meaning and redefine it in instrumental terms. When violence does not look or feel like "war," states and their armed forces have greater power to reshape the meaning of war to suit their political goals. Moreover, they must define war in terms of compelling political goals if they have any hope of rallying members of the military and citizens behind a common cause without provoking the kind of backlash drone strikes have attracted thus far. If war loses its existential character and that character is not replaced by political goals that feel important, war will seem meaningless and therefore illegitimate.

Morality and Selectivity

Mediation's victory over risk not only applies to the risk avoidance associated with force protection but also to the displacement of risk onto specific enemies according to the logic of individuation. Greater control over risk in offensive operations comes largely from the ability to take the lives of specific individuals on the opposing side while sparing others. Drones can displace risk in a calculated and highly individualized manner. They can operate for longer than most manned aircraft, with many existing armed UAVs capable of flying for around twenty-four hours and future UAVs designed

---

47. Coker, *Waging War Without Warriors?*, 71.

to operate for longer still.[48] This makes it possible to track potential targets for extended periods, and even to monitor their "patterns of life" (habits associated with known threats) to detect enemy fighters.[49] Once a target is acquired, UAVs deliver small strikes with low-yield munitions to kill specific individuals or small groups of people while limiting the extent of destruction. Drones have come to prominence in, and helped to create, an era of targeted killing, an era in which individual enemy combatants are marked for death and attacked. Asserting control over risk in the offensive context is therefore closely linked to increased individuation of enemies.

Individuation is not only evident in how drones have been used to target specific enemies but also in how drones have been used to redefine the scope of permissible violence. Supporters of these weapons platforms defend them on the grounds that drones will introduce more humane warfare in which civilians are largely spared from the horrors of fighting. UAVs are contrasted with earlier bombers designed for conventional warfare, which devastated huge areas in their attempts to hit military targets.[50] Drones' capacities for limiting violence to those who are clearly hostile is central to what makes them attractive. Policymakers are able to rationalize attacks to domestic and foreign audiences by claiming that they are employed with the highest level of care for civilians and inflict casualties far short of what would qualify as conventional war. Establishing the normative desirability of drones depends heavily on showing that the risk-management capacities these weapons platforms introduce will ultimately increase individuation to such an extent that only the most important enemies will be killed and that innocent people will be unscathed. This puts drones beyond conventional war and suggests that the instruments of conventional war are destructive to the point of being illegitimate.

Although it is difficult to predict how drone innovation will proceed, the current trajectory suggests that future variants, including those operating on land and at sea, will emphasize precision and the ability to selectively engage enemy fighters. Most existing prototypes evince the same trend of improving precision, which is to say, greater individuation of targets, as the UAVs that are currently in service. Even more importantly, there is good reason to think that political and strategic interests will compel states to continue along this path. The promise of heightened precision that is used to justify drones gen-

48. Woods, *Sudden Justice*.

49. François-Xavier Trégan, "US Drone Strikes in Yemen Cast a Long Shadow over Life on the Ground," *The Guardian*, July 23, 2013, https://www.theguardian.com/world/2013/jul/23/yemen-civilians-drone-strikes-battleground

50. Schulzke, *The Morality of Drone Warfare*.

erates higher expectations for the precision of future operations and increases the public's dissatisfaction when promises are broken.[51] Each generation of drones will need to continue satisfying expectations of increased precision, otherwise support for drone operations could evaporate. Heavily armed drones could become a liability in any case because drones are susceptible to technical faults and hacking.

The future will likely bring drones with higher degrees of autonomy, to the point that drones will be able to identify and attack targets without human guidance. With existing artificial intelligence systems already able to perform delicate maneuvers in flight, it is only a matter of time before drones gain the power of independent targeting (though this may come with human supervision over the choices). There is a precedent for this in the close in weapons systems (CIWS) that have been mounted on ships or stationed along militarized borders for decades. These machines can select and attack targets independently, demonstrating that the will to develop more autonomous machines and the ability to do so already exist. And the next generation of drones with greater autonomy is already taking shape. For example, the US Navy's X-47B UAV, which can take off and land on an aircraft carrier and refuel in flight without help from a human pilot. The Russian army hopes to transform its Armata T-14 into a fully autonomous vehicle at some point in the future and generally shows an interest in automating combat tasks.[52] And Israel is researching several different models of autonomous patrolling and supply vehicles.[53]

The core normative disagreement surrounding autonomous drones is whether they would be able to correctly identify targets. Their critics contend that artificial intelligence systems lack the sophistication to judge innocence and guilt. Machines cannot read contextual cues, understand local norms, or form intuitive evaluations. By contrast, defenders of autonomous drones argue that these machines will be far better at accurately selecting targets than human combatants. Lacking the emotions and cognitive limitations that cloud human soldiers' judgment, and having the ability to fight without concern for self-preservation, autonomous drones may be able to focus violence narrowly at enemy combatants. This is an empirical disagree-

51. James Igoe Walsh, "Precision Weapons, Civilian Casualties, and Support for the Use of Force," *Political Psychology* 36, no. 5 (2015): 507–23.

52. Kris Osborn, "Here's How the US's New Abrams Tank Matches Up Against Russia's Next-Generation T-14 Armata Tank," *Business Insider*, October 25, 2017, https://www.businessinsider.com/heres-how-the-uss-new-abrams-stacks-up-against-russias-t-14-tank-2017-10

53. https://www.timesofisrael.com/unmanned-subs-and-sniper-drones-israel-unveils-its-weapons-of-the-future/

ment that cannot be settled in the immediate future. We will not be able to pass final judgment until we actually see autonomous drones in action. However, in the absence of a conclusive answer, it is at least possible to say that the permissibility of autonomous drones hinges on this issue of whether drones will be able to narrow the application of violence to specific individuals. To satisfy their defenders' predictions of war with fewer civilian casualties, autonomous systems must increase control over the risks associated with fighting, and in particular increasingly individuate war by rigorously selecting targets and avoiding mistaken attacks on civilians. This is revealing, as it shows that target individuation is becoming such an entrenched norm, and so closely associated with drones, that it is one of the essential prerequisites drones have to satisfy to achieve legitimacy.

The Extended Battlespace

Since Hugo Grotius and Emer de Vattel brought just war theory away from its Catholic roots and toward a more secular trajectory, the just war tradition has been premised on regulating violence between Westphalian states that have a monopoly on the use of force within their own territories. The supreme *jus ad bellum* offense was aggression, violating another state's territory. Until the end of the twentieth century, war was only recognized as being legitimate when used to defend a state's territory. The emphasis on territorial defense is weakening now with the rise of responsibility to protect and humanitarian interventions, yet territorial defense remains the most widely agreed-upon basis for fighting and is upheld by the UN Charter.[54] This gives geography the key role when it comes to deciding when the *jus ad bellum* criterion of just cause is satisfied. Acts of aggression and defense only exist when we have some sense of where to draw territorial boundaries. Other *jus ad bellum* criteria are likewise grounded in geography. For example, legitimate authority to declare wars is treated as belonging to the sovereign, whose status comes from the political and military domination of a particular territory.

Jus in bello considerations that pertain during war also depend on geographical assumptions. The principle of discrimination (which forbids intentional or reckless violence against civilians or their property) and principle of proportionality (which urges minimal use of force and prohibits violence that is excessive for achieving its goal) presuppose a territorial division

54. http://www.un.org/en/charter-united-nations/

between geographies of war, in which these principles and the laws of war apply, and geographies of peace that are regulated by domestic laws. As Münkler observes, "The territorial demarcation makes it possible to distinguish clearly between war and peace; precise boundaries are the prerequisite for there to be no intermediate status between war and peace."[55] Different normative and legal paradigms associated with war and domestic law enforcement are grounded in a geographical difference and therefore are threatened by anything that interferes with efforts to draw such a distinction. This is precisely what drones do.

UAVs have upset the geographical underpinnings of war by extending the battlespace worldwide. It was once possible to determine with relative ease where wars were being fought by looking at where combatants were actively engaged. There were, of course, challenges associated with deciding exactly when domestic conflicts qualified as wars or what number of casualties had to be met for violence to become war, but even then, the territory in question could usually be demarcated. Drones raise two distinct geographical/normative challenges. First, from a *jus ad bellum* standpoint, they make it more difficult to determine when wars are taking place and to constrain wars according to traditional normative categories. Drone strikes are violent, yet they fall far short of what we have come to expect from acts of war. Because the attacks are directed at high-profile terrorists and insurgents, rather than at territorial acquisition, they are not aggressive in the sense that would normally qualify as violating the principle of just cause. In fact, they are often rationalized as being essential for promoting the more expansive sense of just cause supported by those who think that human security should be grounds for war just as territorial security is. Adding further complexity to an already labyrinthine issue is that national governments sometimes grant permission for the attacks within enclaves where they are unable to operate (as in the case of Pakistan's South Waziristan),[56] which makes drone strikes seem more like a kind of foreign involvement in domestic law enforcement. The result is that UAV strikes do not fit into conventional war or law enforcement normative/legal paradigms. One could say that they occupy a gray area, but even this would be too generous when their use does not simply come in a place of indistinction between these paradigms but rather challenges the applicability of the paradigms in any future conflict.

Second, drones upset the geography of war by introducing the revolutionary possibility of participating in combat while remaining safely within

55. Münkler, *The New Wars*, 38.

56. Woods, *Sudden Justice*, 234.

what would ordinarily be considered spaces of domestic life. Most attacks by US and UK forces are controlled by pilots who live and work within those countries. When the norms of war are geocentric, geographical transgressions like this become normative transgressions as well. Drone operators live in uncontested areas in which killing qualifies as murder according to the law enforcement paradigm, but they are able to act by proxy in areas where killing is justifiable and expected according to the war paradigm. The disjuncture is underscored by how the pilots return to civilian life between shifts. Martin reports that "It could be disorienting, a real disconnect, to live simultaneously in two such different worlds."[57] If we seek to identify where wars are taking place by looking at the participants, we must conclude that the wars in Iraq, Afghanistan, Yemen, Syria, Libya, Somalia, and Gaza (among other places) are also taking place in the United States, United Kingdom, and Israel (among other places) by virtue of the fact that military personnel in those countries are actively participating in hostilities. War is global, but not treated as such, so normative double standards—especially opportunistic use of law enforcement and war norms—are rampant.

The effects of deterritorialization are fairly modest at present, owing to the relatively limited use of drones and their deployment in counterinsurgency and counterterrorism operations. Nevertheless, the normative change underway is profound. As drones become more widespread, opposing combatants will be forced to enter the drone user's territory to find legitimate military targets. This will make it impossible to retain the sanctity of domestic spaces as places beyond war. General Stanley McChrystal speculated about the possibility of Islamic State carrying out attacks against drone operators stationed in Western countries, saying that "Anything we use that's part of that [defense infrastructure] becomes, in my view, a fair target for our enemies, and we have got to consider that."[58] He is right in seeing that the logic of international terrorism and the reliance on drones point to a future in which enemy fighters' only chance of victory is to attack military personnel at home. The same incentives will be at work in wars between more evenly matched opponents, such as a war between two states using drones. As violent non-state actors make greater use of drones there is also the chance that they too could bring war into domestic spaces, such as if those fighting in Syria and Iraq were to pilot vehicles from across the border in Jordan or Leb-

57. Martin and Sasser, *Predator*, 45.

58. Richard Norton-Taylor and Alice Ross, "RAF Base May Be Legitimate Target for Isis, Says Ex-NATO Commander," *The Guardian*, November 25, 2015, http://www.theguardian.com/uk-news/2015/nov/25/raf-base-may-be-legitimate-target-isis-ex-nato-commander

anon. To the extent that either side relies on drones to fight from a distance, the other will have grounds for fighting in the domestic spaces where drone pilots live and work.

Drone Warriors?

Drones are changing the military profession by encroaching on roles that previously required physical presence on the battlefield. The shift is occurring slowly due to military conservatism, and meets considerable resistance from those who are invested in more traditional methods of fighting associated with conventional war, but it is well under way and has passed the point of no return. The US Air Force showed little interest in using drones for combat operations until the Central Intelligence Agency armed its own Predator UAVs and began launching targeted attacks. Once this reluctance was overcome, the Air Force and Joint Special Operations Command became avid users of these machines. Ground and naval drones have not been deployed in large numbers, but the proliferation of prototypes has set the conditions for a similar rapid escalation across these terrains in future conflicts.

The effects of drones on the military profession in general can be extrapolated from their impact on the US Air Force. Throughout the Cold War, the Air Force's identity was built around the romanticized fighter pilots and the Strategic Air Command's immense nuclear power. Those two domains came to define its culture, and led to the dominance of personnel from those two areas of service, along with their culture and institutions. The Air Force faced a threat to its prestige and relevance when many important UAV missions came under CIA control.[59] It had to compete by investing more heavily in UAVs. This is producing a cultural and institutional shift. Now pilots are at risk of being marginalized in favor of drone operators who wage wars from inside control trailers. The future of manned fighter aircraft and strategic bombers is in doubt, as is the dominance of the associated personnel and culture.[60] As air forces around the world make the switch to UAVs they are experiencing a similar disruption of professional identity and contemplating a future without any humans in the cockpit. The spread of drones into other domains will extend this process to the point that it will be possible for

59. Woods, *Sudden Justice.*

60. Rachel Martin, "Drone Pilots: The Future Of Aerial Warfare," NPR, November 29, 2011, https://www.npr.org/2011/11/29/142858358/drone-pilots-the-future-of-aerial-warfare

machines to perform the dangerous roles that have traditionally distinguished members of the military.

Drones undermine the status of those military personnel who once enjoyed the greatest prestige because they introduce a new type of combatant: the technical expert. The US Air Force initially tried to employ conventional aircraft pilots as UAV operators, but discovered that they imported behaviors and biases that are inappropriate for the role. Personnel with no prior flight experience turned out to be more suitable, leading the Air Force to seek recruits who might have once been marginalized and relegated to supporting roles. Thus, investment in UAVs is not just a matter of switching pilots from one type of aircraft to another. It instead depends on creating an entirely new identity that exists apart from, and in opposition to, that of other pilots.

Existing UAVs are relatively unsophisticated machines when it comes to their command and control. They are guided with the same kind of autopilot systems used in manned aircraft and are unable to do anything more than orbit above a target area when they lose direct input from their controllers. This lack of independence makes drones vulnerable to bad weather, hacking, and interference with communications. The next generation of systems will likely rely far more heavily on artificial intelligence, especially when it comes to movement controls, to prevent loss when the link between machine and operator is severed. As drones grow in autonomy, traditional military roles will come under greater threat, both because technical specialization will increase and because human soldiers may play a shrinking part in combat. It is even possible to realistically envision a future in which the defining moment of combat—the killing of an enemy soldier—is entirely automated or in which military personnel only perform a supervisory role. Such a change would be the most radical step toward instrumental warfare imaginable, completely removing the human element of war and reducing it to the strategic application of violence without any pretense of higher existential meaning.

Increasing reliance on drones will change the military profession far beyond those personnel who are directly involved with unmanned systems. For the military, norms are closely linked to professional identity. Soldiers are trained to follow distinct codes of conduct because their roles are profoundly different from those of the civilian world, and the professional identity is linked to an existential conception of war. Fighting and killing are ordinarily considered immoral, but are expected of soldiers. Members of the military usually live apart from civilians, occupying their own insular world with unique practices and identities. Many of the world's leading militaries maintain their own sets of virtues that soldiers are supposed to embody, a classical conception of "the warrior ethos."[61] These are typically grounded in traditional military roles performed by soldiers who directly confronted their enemies on the battlefield. Moreover, soldiers are taught to venerate the classical heroes as exemplars of virtue, and to model themselves on the veterans of past wars who distinguished themselves by taking enormous personal risks in close combat.[62] Such guiding figures of existential war will seem anachronistic in an era when many soldiers fight by proxy. Those warrior archetypes may even need to be explicitly repudiated to preserve the legitimacy of mediated warfare that so blatantly contradicts the qualities that have made past wars seem heroic.

Drones threaten traditional conceptions of virtue, and in doing so undermine the systems of military ethics that have developed around them. As Münkler points out, "The classical army ethos, which for a long time was probably the most reliable obstacle to the conversion of fighting into massacre, has been supplanted by a combination of technical precision and legal control."[63] Robert Sparrow likewise contends that military virtues will lose their meaning in drone operations.[64] Jesse Kirkpatrick has attempted to show that virtues can remain meaningful, albeit in a different form. The physical courage of facing enemies on the battlefield, for example, may be replaced by the moral courage of ensuring that weapons are used responsibly and that misconduct is reported. However, this effort to save virtues comes at the expense of completely changing their meaning. The result of losing or redefining virtues is the same: the collapse of the military's existing professional identity and its norms. This does not mean that force will be used unethically in the future. On the contrary, there is good reason to think that drones can be used with greater restraint by affording combatants the freedom to make targeting decisions without having to worry about their personal safety.[65] New guiding principles may be found, but these will need to come from different sources and comprise a different type of military culture.

Critics often argue that drones' assault on traditional military virtue ethics makes them morally objectionable. Münkler and Sparrow both use this

---

61. Paul Robinson, Nigel De Lee, and Don Carrick, eds., *Ethics Education in the Military* (Hampshire: Ashgate, 2008).

62. Shannon E. French, *The Code of the Warrior: Exploring Warrior Values Past and Present* (New York: Rowman & Littlefield, 2003).

63. Münkler, *The New Wars*, 132.

64. Robert Sparrow, "War without Virtue?" In *Killing by Remote Control*, ed. Bradley Jay Strawser New York: Oxford University Press.

65. Schulzke, *The Morality of Drone Warfare.*

line of reasoning, venerating traditionalism and bemoaning its subversion by new technologies. However, it is important to point out that the normative implications of changes in the military profession are more complex than this and that many traditional military values are morally objectionable.[66] The existential narrative of war as the great test of patriotism and selfless dedication imbues violence with a special normative status and shifts attention away from the harsh realities of conflict. Drones upset illusions about war being fair and render heroic narratives about war as an ennobling experience untenable.

It is worth remembering how dangerous the existential conception of war can be, since it celebrates conflict as being an inherently meaningful, redemptive experience that allows participants to become heroes. Discrediting this narrative may make war less attractive. Martin Shaw aptly notes that a new culture may be forming. "The glorification of military institutions and values was replaced by the cult of the weapon-system: traditional militarism was replaced by 'armament culture.'"[67] What remains to be seen is whether a culture built around the instrumental application of violence will be more or less destructive than one built around a myth of military glory. This is an open question, but at present we have the opportunity to deliberately construct the emerging military culture with stronger moral sensibilities than those that informed the military culture that is in decline. The fact that militaries are showing unprecedented concern with ethics education suggests that greater instrumentalism may be morally advantageous.[68]

Conclusion

Future wars will no doubt take myriad different forms depending on the participants' goals, structures, and relative power. However, belligerents in any type of conflict will likely be attracted to drones because of their potential to simultaneously satisfy the demands of force protection and greater precision. It is difficult to imagine a conflict scenario in which these goals would not be operative, so it is reasonable to assume that drones will have

66. Timothy L. Challans, *Awakening Warrior: Revolution in the Ethics of Warfare* (Albany: State University of New York Press, 2007).

67. Martin Shaw, *The New Western Way of War: Risk-Transfer War and Its Crisis in Iraq* (Malden, MA: Polity Press, 2005), 37.

68. Robinson, De Lee, and Carrick, *Ethics Education in the Military*; James Connelly, Don Carrick, and Paul Robinson, eds., *Ethics Education for Irregular Warfare* (Burlington, VT: Ashgate, 2009).

enduring appeal. As drones become more integral to armed forces around the world, they will mediate combat on a greater scale to grant users unprecedented risk-management capabilities. Drone operators may become vulnerable to enemy attacks (from enemy drones, or from more conventional means), but the dangers to them will not come from any risk to the machines they control. The fate of human operators and the machines will remain sharply divided, with attacks against one having no immediate effect on the other. The reliance on physical mediation will push war away from the existential understanding that has historically prevailed and toward an instrumental view, because the characteristics that once sustained the existential perspective are untenable under the pace of technological change. That is to say, war is apt to continue, but the instrumental political function of war will be more central. The celebration of combat as an edifying experience is anachronistic to the extent that combat is being conducted through drones.

Drones' disruption of the geocentric norms of just war theory and international law should prompt greater attention to the normative constraints that just war theorists and jurists have sought to impose. It is particularly important to recognize that the norms inherited from the Westphalian era have lost much of their power because of the many ways in which drones extend conflict beyond the boundaries of conventional war. Changes to the military profession make normative challenges all the more difficult by threatening the systems of military ethics that have been used to guide soldiers. Again, the instrumental character of war is set to triumph. The norms and professional identities that are under threat are heavily invested in more traditional understandings of war grounded in an outdated warrior ethos. Those values are unsuitable for drone warfare and must either be completely revised or reinterpreted in such a radical way that their content will be fundamentally different.[69] Already the conscience of drone strikes comes not from the warriors themselves but from lawyers and bureaucrats who represent the instrumental understanding of war. Control is centralized at higher levels, and easier oversight encourages rule following over soldiers' exercising their own judgment. This will have mixed costs and benefits. Change brings normative uncertainty and disruption, yet reducing war's attractiveness is an important first step toward overcoming ideologies that promote militarism and cloak the horrors of fighting with uplifting narratives. In other words, technocratic warfare is profoundly unappealing, but that is probably a good thing.

---

69. For more about this see Marcus Schulzke, "Rethinking Military Virtue Ethics in an Age of Unmanned Weapons." *Journal of Military Ethics* 15, no. 3 (2016): 187–204.

2 • The Digital Battlespace

Subduing the Enemy without a Fight

Cyberattacks are extraordinarily versatile. They have been launched to support conventional military operations, such as during Russia's attack on Georgia in 2008, in stand-alone offensives, as in the Stuxnet attacks on an Iranian uranium enrichment facility, and for nonviolent espionage.[1] They have been used by states and nonstate actors alike, and against targets that range from small nonstate actors to the world's most powerful countries. They can be deployed in pursuit of a vast range of objectives, from stealing information, to sabotaging physical infrastructure, to influencing elections. This versatility makes cyberweapons attractive for all types of violent actors from the smallest criminal enterprises and insurgencies to the most powerful states. Like drones, cyberweapons can precisely calibrate attacks according to the logic of individuation, so that they strike specific enemies without causing collateral damage. Also like drones, they offer improved risk management through mediation, although cyberweapons take this mediation one step further by not only overcoming geographical constraints but also concealing an attack and its source.

New technologies tend to improve certain risk-management techniques while also introducing new risks that those technologies could themselves fail, with potentially catastrophic results. Coker notes that "As we introduce more sophisticated technology, new risks proliferate at an exponential rate."[2] Nowhere is this clearer than from cyberweapons. They are uniquely risky because they attack from within, transforming digital infrastructure into a liability. They may undercut the efficacy of other revolutionary military tech-

1. Kim Zetter, *Countdown to Zero Day: Stuxnet and the Launch of the World's First Digital Weapon* (New York: Crown, 2011).

2. Christopher Coker, *Waging War Without Warriors?* (Boulder: Lynne Rienner, 2002), 5.

nologies, such as by disabling or seizing control of drones. One of the nightmare scenarios for the future of international security, which has been considered in strategic theory as well as in popular culture, is the possibility that drones could be seized and used against their owners. Worse still, cyberattacks have a boomerang effect. States that launch them provide a guide for opponents' own attacks, even giving them code that may be copied and incorporated into new digital weapons. Sadly, this is already evident from the WannaCry worm, which spread to over 200,000 computers in 2017 with the help of exploits discovered by the US National Security Agency.[3] This kind of risk boomerang also affects defensive technologies, such as encryption. Fears that the NSA builds back doors into its advanced encryption technologies have been a major factor in preventing their adoption over less sophisticated methods.[4]

Because they transform an enemy's strengths into weaknesses, cyberweapons have a parasitic character. The most sophisticated attacks even make security systems into accomplices. The Stuxnet worm, for example, concealed itself by monitoring the normal functioning of the Iranian uranium enrichment centrifuges, then using that data to issue false reports that the systems were operating under the same parameters once the attacks began.[5] More serious than any physical damage, cyberweapons cause psychological trauma by spreading fear and uncertainty. They attack perceptions far more effectively than they have attacked any physical infrastructure. The practice of striking from within undercuts trust in machines and promotes fear about when and how adversaries could strike.[6] High-profile attacks are often followed by panic about the future of cybersecurity and lost faith in the institutions that are supposed to keep us safe. The potential for this kind of perceptual assault is evident from how thoroughly Western governments were shaken by the threat of Russian interference in their elections following that country's theft of damaging materials during the 2016 US presidential election.[7]

Deciding to include cyberattacks in an analysis of war is apt to provoke

3. Lily Hay Newman, "The Leaked NSA Spy Tool That Hacked the World," *Wired*, March 7, 2018, https://www.wired.com/story/eternalblue-leaked-nsa-spy-tool-hacked-world/

4. Bruce Schneier, *Data and Goliath: The Hidden Battles to Collect Your Data and Control Your World* (New York: W. W. Norton, 2016).

5. Zetter, *Countdown to Zero Day.*

6. Paranoia and confusion can have a serious impact on civilian government officials and members of the military, as shown in Fred Kaplan, *Dark Territory: The Secret History of Cyber War* (New York: Simon & Schuster, 2016).

7. Luke Harding, "Top Democrat's Emails Hacked by Russia after Aide Made Typo, Investigation Finds," *The Guardian*, December 14, 2016, https://www.theguardian.com/us-news/2016/dec/14/dnc-hillary-clinton-emails-hacked-russia-aide-typo-investigation-finds

objections from those who contend that cyberwarfare is a vacuous or illusory concept. Rid argues that "Cyber war is a highly problematic, even a dangerous, concept. . . . No stand-alone cyber offense on record meets these criteria, so 'cyber war' remains a metaphor for the time being."[8] Rid may be right in thinking that cyberwar is unlike conventional war, or even many types of unconventional war, but I contend that this really demonstrates the inadequacy of prevailing conceptions of what qualifies as war. The innovations I discuss throughout this book show that the line between war and peace is extraordinarily difficult to identify with any precision in an era when political violence and nonviolent coercive tools are continually used in ways that do not qualify as conventional warfare, but that are also far from peaceful. The innovations show coercion operating beyond the boundaries of "war" and "peace" as these terms are commonly used. They also evince political incentives for operating within that gray area and opportunistically using norms appropriate to one domain or the other.

I argue that one of the foremost reasons why states and nonstate actors are drawn to cyberwarfare is that it is difficult to conceptualize in a simple war vs. peace binary. Cyberattacks provide a means of exercising coercion while minimizing the risk of reprisal, whether from the adversary or from domestic and international stakeholders who oppose taking hostile action. In other words, cyberattacks are often attractive coercive instruments because of an uncertainty about the concept of warfare. The world's leading military powers have helped to create this uncertainty by pioneering many tools of cyberwar and frequently using them in ways that obscure the meaning of "war." Future conventional military operations will likely be coupled with cyberattacks, as in the case of Russia's invasion of Georgia. However, it is outside the conventional war domain that the unique appeal of cyberattacks as low-risk coercive tools are most evident.

Disentangling Cyberwarfare

It is difficult to write a brief history of the cyberattacks launched by states because key events in this history are often obscured by uncertainty and misinformation. According to some former CIA operatives, the first lethal cyberattack was launched in 1982, when the sabotage of the Urengoy-Surgut-Chelyabinsk natural gas pipeline in Russia caused a massive explo-

8. Thomas Rid, *Cyber War Will Not Take Place* (New York: Oxford University Press, 2013), 37.

sion.[9] By other accounts, this event was simply an accident that was opportunistically claimed as a successful American attack.[10] It is revealing that one of the prime candidates for the first act of cyberwarfare is shrouded in doubt. When it comes to cyberwarfare, uncertainty reigns supreme.

Although we do not know whether cyberattacks have been lethal, it is clear that they can cause physical damage.[11] The Stuxnet worm that was identified in 2010 was designed to destroy the uranium enrichment centrifuges at Iran's nuclear facility at Natanz by making them spin irregularly. The worm may have destroyed one to two thousand of ten thousand centrifuges before it was detected.[12] Moreover, it caused irreparable perceptual damage. Scientists were dismissed during the blind search to discover who was responsible for the failures; enrichment was halted multiple times; confidence in computer systems was shaken. It was a case of unattributed damage engendering misperceptions and mistrust. It was only years after its initial detection that Stuxnet was definitively linked to a joint US-Israeli venture called Operation Olympic Games. Worry about undetected infections has had an enduring legacy. Some Iranian officials blamed exploding natural gas and oil pipelines during July and August 2016 on cyberattacks even though there was little evidence to support that narrative.[13]

Although the issue of whether cyberattacks can cause serious physical harm or casualties is important, it should not detract from the more urgent problem of *what else they might be able to do*. It is more profitable to think about how cyberattacks can inflict destruction in an indirect sense, without actually causing physical harm, as this is the source of their unique advantages over other coercive instruments. Rather than seeing the lack of lethal attacks and limited number of attacks causing physical destruction as evidence that cyberattacks have limited efficacy or that they cannot qualify as acts of war, we should take this as evidence that states have found it more profitable to use these as an alternative means of coercion beyond war and peace. The United States and Israel had the power to bomb the Natanz facility from the air. They

---

9. Alec Russell, "CIA Plot Led to a Huge Blast in Siberian Gas Pipeline," *The Telegraph*, February 28, (2004), http://www.telegraph.co.uk/news/worldnews/northamerica/usa/1455559/CIA-plot-led-to-huge-blast-in-Siberian-gas-pipeline.html

10. Rid, *Cyber War Will Not Take Place*, 5–7.

11. Although Rid is generally skeptical of the concept of "cyberwar" and of purported cases of cyberattacks causing physical harm, he does acknowledge that attacks causing physical harm are plausible.

12. Zetter, *Countdown to Zero Day*.

13. Kay Armin Serjoie, "Iran Investigates If Series of Oil Industry Accidents Were Caused by Cyber Attack," *Time*, August 12, 2016, http://time.com/4450433/iran-investigates-if-series-of-oil-industry-accidents-were-caused-by-cyber-attack/

could have even attempted a ground attack. They declined those options. Two of the most powerful armed forces in the world faced a weaker adversary and decided that the best attack was one they could launch without overt aggression. In doing so, they set a precedent for stepping beyond the boundaries of the Westphalian system and conventional warfare.

Sun Tzu's core theoretical contribution is that war not only exists in the moment of contact between two opposing forces but also in the preparatory effort and that the highest skill is to manage preparations so effectively that it becomes possible to secure victory without having to fight. As he says, "to fight and conquer in all your battles is not supreme excellence; supreme excellence consists in breaking the enemy's resistance without fighting."[14] He likewise advises that "the skillful leader subdues the enemy's troops without any fighting; he captures their cities without laying siege to them; he overthrows their kingdom without lengthy operations in the field."[15] From this perspective, avoiding physical destruction is not a way of avoiding war but rather a higher mode of warfare that skillful commanders should aspire to. Cyberattacks may cause physical destruction, but in most instances they are compelling precisely because they do not require this. Even when physical destruction is necessary, it is often better to hide it to create the impression that no attack has taken place.

Russian operations illustrate just how far indirect attacks may go in achieving strategic objectives without lethality. Vladimir Putin's program of reasserting Russia's status as a global power has set it on a course for conflict with NATO, but he has wisely avoided conventional hostilities. Russian hackers manipulated the 2016 US presidential election by releasing damaging e-mails and creating defamatory social media content. Its cyber campaign reveals an effort to change the future of American governance without direct physical confrontation. We will never know what would have happened without Russian involvement in the election, but it is plausible that it played a role in securing Trump's victory and that it therefore shaped subsequent US operations. Of particular importance is that election interference may have helped Russian operations in Ukraine and Syria by reducing the likelihood of US opposition in those theaters. What is clear is that Russian influence damaged American prestige and fueled fears of a resurgent Russian cyber threat that continues to the present.[16] Russia dealt serious blows to confi-

---

14. Sun Tzu, *The Art of War* (Los Angeles: Enhanced Media, 2017), 13.

15. Sun Tzu, *The Art of War*, 13.

16. David Chazan, "France Blocks 24,000 Cyber Attacks Amid Fears That Russia May Try to Influence French Presidential Election," *The Telegraph*, January 8, 2017, http://www.telegraph.co.uk/news/2017/01/08/france-blocks-24000-cyber-attacks-amid-fears-russia-may-

dence in democratic governance and the future of NATO without firing a shot. This was not war in a conventional sense, but it perfectly captures the approach Sun Tzu advocates. It likewise shows how offensive operations are extending beyond the boundaries the Westphalian system is premised on. Such an interference in domestic politics is contrary to the spirit of sovereign authority over domestic affairs and does not conform to the norms of conflict between states.

Sun Tzu also advises that when it is not possible to avoid fighting, a skillful commander manipulates the structural conditions of the battle to guarantee success. The goal is to prevent combat from being a fair fight by ensuring that the enemy will fail. That is to say, Sun Tzu calls for war to be asymmetric. Any symmetric arrangement of forces would mean that the decisive advantage has been lost, that victory is not assured, and that a commander is making a foolish gamble. Cyberattacks are attractive tools for achieving the kind of structural advantage Sun Tzu calls for. Even if they do not cause physical destruction, they may erode the enemy's capacities as part of a broader campaign. During the 2008 Georgia War, Russian distributed denial of service (DDoS) attacks against Georgian government computers and media outlets hindered internal and foreign communication, a critical move when Georgia's only hope for victory lay in securing foreign assistance as quickly as possible. These attacks were far from spectacular, but for a first foray into combined arms cyberwarfare they show a strong grasp of how strikes against communications systems could help to isolate adversaries, thereby preventing external disruption of an asymmetric balance of conventional forces that guaranteed a Russian victory. Ongoing strikes against Ukraine while war rages there further indicate that Russia considers cyber operations to be an important offensive capacity and that it is continuing to refine them.

Mediation and Risk Displacement

The previous examples display risk-management strategies concerned with accomplishing objectives with the lowest possible risk to the attacker's forces and political standing, while also ensuring that any open hostilities will be asymmetric. As with drones, this control over risk is asserted via mediation. Cyberattacks are necessarily mediated. They are launched

try/; Melissa Eddy, "After a Cyberattack, Germany Fears Election Disruption," *New York Times*, December 8, 2016, https://www.nytimes.com/2016/12/08/world/europe/germany-russia-hacking.html

through computer systems and against computer systems, usually with intermediary systems concealing an attack's source and intent. Mediation is not simply a means to an end but also an end in itself; it transforms the enemy's computers into weapons while also disabling them or interfering with their normal functions. As Rid says, "code-caused violence is indirect: it has to 'weaponize' the target system in order to turn it into a weapon. Code doesn't have its own force or energy. . . . Code-caused destruction is therefore parasitic on the target."[17] An enemy's systems facilitate an attack, not only making it possible in the first place but also potentially allowing the effects to ripple outward when inadequate safety precautions are in place to constrain the effects. The target's networks, which may ordinarily be a strength, become a liability. The more the target relies on computers and the more those are integrated, the more vulnerable the target becomes.

Mediation allows the attacker to strike while facing little risk of retaliation or even detection. In almost all cases, the ultimate target is a piece of software or hardware, not a person or group of people. Avoiding human casualties is advantageous, as this minimizes the chance of detection and undercuts the normative case for a lethal response. With cyberattacks taking place through computer systems from inception all the way to impact, there is often little evidence that an attack has taken place. The long process of detecting and investigating attacks further ensures that time and the bureaucratic politics of forensic research will reduce an attack's immediacy and diminish the sense that cyberwarfare is occurring in real time. By the time an attack is uncovered and the culprit identified, it may be too late for the victim to make a credible case for a lethal response.

The parasitic character of cyberweapons leads to yet another crucial strategic benefit. Sun Tzu reflects on how quickly military strength is degraded by prolonged fighting and warns that the effort to replace losses and supplies will outstrip any state's capacities. Pyrrhic victories sap military strength, making it difficult to translate success on the battlefield into political success. Great commanders from Hannibal to Robert E. Lee have fallen victim to this problem, watching their armies deteriorate from victory after victory and being unable to accomplish their overall strategic mission. Sun Tzu advises commanders to offset the costs of fighting by taking as much as possible from the opponent. "Hence a wise general makes a point of foraging on the enemy. One cartload of the enemy's provisions is equivalent to twenty of one's own."[18] Appropriating material from an enemy has the dual benefits of enriching one's own forces

17. Rid, *Cyber War Will Not Take Place*, 13.
18. Sun Tzu, *The Art of War*, 12.

while also taking resources away from the enemy. Sun Tzu applies the same logic when dealing with military equipment and even with soldiers themselves. "Therefore in chariot fighting, when ten or more chariots have been taken, those should be rewarded who took the first. Our own flags should be substituted for those of the enemy, and the chariots mingled and used in conjunction with ours. The captured soldiers should be kindly treated and kept."[19] When it comes to conventional warfare, captured equipment and supplies are sometimes useful, but captured soldiers can rarely be redeployed into effective fighting units. The German and Soviet militaries both tried this during the Second World War, with little success. Sun Tzu's advice seems a bit naïve in this respect, since the possibility of converting soldiers assumes an extremely weak attachment to their cause and their comrades.

Cyberwarfare makes it much easier to realize the dream of living parasitically on others' strengths by shifting the allegiance of compromised computer systems. Stuxnet transformed the computers meant to regulate nuclear enrichment into tools for destroying centrifuges. Related worms, such as Gauss, Duqu, and Flame, transformed government computers into surveillance devices.[20] Cyberweapons take this resource theft one step beyond Sun Tzu; they not only draw strength from the adversary's strengths but also from a broad range of third-party networked devices. This is evident from the practice of constructing botnets: assemblages of hijacked computers that can be directed by the attacker, often without the computers' owners being aware. Botnets are typically used to launch DDoS attacks that temporarily overwhelm an adversary's computers, rendering them ineffective. Russia's assaults on Georgian systems were carried out with the help of botnets, as were DDoS attacks against Estonia the year before.[21] The infected systems that made up the attacks were scattered around the world, essentially transforming all the digital resources Russian hackers were able to access into tools for launching their attacks.

Botnets make it possible for nonstate actors to dramatically increase their strength. Although it has cultivated an image of populist hacking that depends on collective involvement, much of Anonymous' strength comes from the botnets controlled by its central members.[22] In attacks against

19. Sun Tzu, *The Art of War*, 11.

20. Boldizsár Bencsáth, Gábor Pék, Levente Buttyán, and Mark Felegyhazi. "The Cousins of Stuxnet: Duqu, Flame, and Gauss," *Future Internet* 4, no. 4 (2012): 971–1003.

21. John Markoff, "Before the Gunfire, Cyberattacks," *New York Times*, August 12, 2008, https://www.nytimes.com/2008/08/13/technology/13cyber.html

22. Parmy Olson, *We Are Anonymous: Inside the Hacker World of LulzSec, Anonymous, and the Global Cyber Insurgency* (New York: Little, Brown, 2012).

Amazon, PayPal, Visa, and other corporations, these acted like zombie armies that could be launched against a target with little risk of retaliation. Such a strategy only works in this heavily mediated environment that has made it possible to more perfectly realize Sun Tzu's advice about appropriating resources. As networked devices continue to proliferate and build an "Internet of things," the opportunities for building botnets will continue to grow. Recent attacks have passed through baby monitors, security cameras, mobile phones, doorbells, and smart kitchen appliances, weaponizing these artifacts of everyday life.

Because they are so heavily mediated in their delivery and effects, cyberattacks rarely feel like attacks. Few produce any physical damage and most do not inflict any costs that can be measured in terms of conventional military calculations. Costs and benefits usually take the form of information lost or won, future development plans being thwarted, communication being interrupted, and the erosion of trust. As Rid notes, "Violence administered through cyberspace is less direct in at least four ways: it is less physical, less emotional, less symbolic, and, as a result, less instrumental than more conventional uses of political violence."[23] The lower emotional and symbolic significance is particularly important when it comes to understanding why cyberattacks are perceived so much differently than attacks involving conventional weapons. Cyberattacks provoke the anxiety and paranoia of unexpectedly suffering the failure of one's own computer systems, yet any particular attack lacks the same impact as acts of physical violence. This is especially true when largely invisible cyberattacks, such as Russia's ongoing operations against Ukraine, Georgia, and the Baltic states, are compared against visually spectacular violence of events like the First Gulf War, the invasion of Iraq, or the destruction of Aleppo. Minimizing the appearance of offensive activities to such an extent that it is even doubtful whether cyberattacks constitute a form of warfare has been integral to keeping these attacks out of the news more effectively than is possible with drone strikes. This in turn allows aggressors to shield themselves from public opinion costs they might suffer if they perpetrated lethal operations.

New Domains of Risk

Rasmussen says of risk that "if one prevents a scenario from becoming real, the result will probably be to create new risks, which then rise to the top of

23. Rid, *Cyber War Will Not Take Place*, 12.

the agenda. The theoretical outcome of this process is that risks are infinite because they multiply over time since one can always do more to prevent them from becoming real."[24] For example, the 2003 invasion of Iraq was proffered as an effort to prevent terrorism and the proliferation of weapons of mass destruction, yet destabilizing the Iraqi government created a political vacuum in which new threats could arise.[25] Similarly, a cyberattack deployed against an enemy can easily rage out of control, infecting the computer systems of allies or the actor that initially deployed it. This leads to some challenging risk-management scenarios.

Some cyberattacks foster selective risk acceptance. States wielding them must preserve avenues for attack in enemy computers, so they sometimes have to resist the temptation to patch those same vulnerabilities in allied computers. As Joseph Menn points out, "Spy tools and cyber-weapons rely on vulnerabilities in existing software programs, and these hacks would be much less useful to the government if the flaws were exposed through public warnings. So the more the government spends on offensive techniques, the greater its interest in making sure that security holes in widely used software remain unrepaired."[26] A state mounting a cyberattack could attempt to protect its own computer systems against known exploits to minimize the risk of collateral damage, but that would signal the existence of vulnerabilities and render the weapon ineffective against enemies who quickly develop their own patches. Attacks only succeed to the extent that they are unexpected by friend and foe alike. The risk of a boomerang is particularly high when an attack is launched by a state but may be turned against critical infrastructure that is under civilian control and that may therefore be more difficult for the government to covertly secure. This perverse incentive to preserve certain risks is greater in the cyber domain than it is elsewhere because offensive and defensive advantage are largely achieved via information that can in principal be easily transferred, as opposed to conventional advantages that depend more on material resources.[27]

---

24. Mikkel Vedby Rasmussen, *The Risk Society at War: Terror, Technology and Strategy in the Twenty-First Century* (New York: Cambridge University Press, 2006), 4.

25. David Kilcullen, *Blood Year: Islamic State and the Failures of the War on Terror* (Oxford: Oxford University Press, 2016).

26. Joseph Menn, "SPECIAL REPORT—U.S. Cyberwar Strategy Stokes Fear of Blowback," *Reuters*, May 10, 2013, http://in.reuters.com/article/usa-cyberweapons-idINDEE9490AX20130510?type=economicNews

27. Fred Kaplan finds evidence of American officials concealing vulnerabilities in mainstream operating systems to keep these open for its own purposes. See Kaplan, *Dark Territory*.

Equally serious is that each cyberweapon provides a model for others to emulate. "Using a cyberweapon is equivalent to sharing it—the code, and thus the capability, residing in the weapon is now loose in the virtual ecosystem, where it can be picked up, studied, and reused by the intended target, or anyone else."[28] This makes the deployment of a cyberweapon, especially one introducing "zero day" exploits—exploits that have not been discovered by security professionals—a threat to any computer systems, even those belonging to the actor deploying the cyberweapon. Some critics argue that the United States and Israel acted irresponsibly by using Stuxnet, as this worm created a template for other attacks on industrial control systems and made these available to hackers around the world.[29] The worm's payload was narrowly focused, but the exploits used to deliver it and to avoid detection could be appropriated for other purposes. The modular form of sophisticated worms like Stuxnet magnifies this risk by making it possible to selectively take certain pieces of code. Worms are Frankenstein creations made up of various new and old components pieced together to save time and energy.[30] With each new exploit releasing these pieces of code to the world, it is impossible to launch a credible offense without also offering adversaries weapons for their own arsenals.

Risk-management strategies may have ironic effects not only when it comes to offense but also when actors are attempting to build their defensive capabilities. Managing risks on the defensive can be even more challenging than limiting the effects of aggression. Attackers have a high degree of control over when and how risks are allocated—at least in the first instance—giving them the ability to invest their resources with more certainty about where they will be needed. And some attacks, such as DDoS attacks, can be directed against specific systems with little chance of collateral damage. Defenders, by contrast, are left persistently imagining what form a potential attack might take and must struggle to allocate their finite resources to anticipate and avert myriad potential dangers based on probabilistic assessments. Some defenders may attempt to lure attackers to particular targets, such as honey pots hiding malicious code or misleading information.[31] Drawing attackers toward well-defended targets depends on carefully managing per-

---

28. David Kilcullen, *Out of the Mountains: The Coming Age of the Urban Guerrilla* (New York: Oxford, 2013), 138.

29. Zetter, *Countdown to Zero Day.*

30. Zetter, *Countdown to Zero Day.*

31. Erik Gartzke and Jon R. Lindsay, "Weaving Tangled Webs: Offense, Defense, and Deception in Cyberspace," *Security Studies* 24, no. 2 (2015): 316–48.

ceptions: deceiving potential attackers and carefully guarding the overall defensive strategy from public view.

Hiding Responsibility

Whether offensive or defensive, the risk-management strategies deployed in the cyber realm are largely invisible to the public, thereby providing opportunities for perception management. This invisibility is often essential, as knowledge of computer security protocols is usually a precondition for defending against attacks or bypassing defenses. However, secrecy makes it extraordinarily difficult for members of the general public to gauge the effectiveness of cyberwarfare strategies. As Rasmussen points out, risk naturally occludes efforts to assess policy decisions because the costs and benefits are so often only expressed probabilistically or as counterfactuals. "A risk is a scenario followed by a policy proposal for how to prevent this scenario from becoming real. For this reason, a success cannot be measured with any degree of finality because success depends on creating a reality different from what one feared would happen."[32] Without reliable metrics for gauging success and failure, these concepts lose their utility. This is especially true given public ignorance of cyberwarfare compared to conventional operations. It is extraordinarily difficult to determine when key actors responsible for providing security are actually doing so. In other words, cyberwarfare creates a state of affairs in which the necessary steps for managing risks interfere with government transparency. This makes it difficult to hold government officials accountable for what they do, or fail to do, to provide security in this domain.

Lacking information about how operations are being managed and whether they are successful, ordinary citizens are disempowered and have limited oversight capacities. Members of the public are then left with little to inform their evaluations of whether policymakers and cybersecurity experts are working effectively. They cannot even look for the usual signs of conflict, such as dead and wounded bodies, for a more concrete sense of what is happening. People lack a baseline for what is normal and are apt to only hear about cyberattacks when these are especially dangerous, which increases the salience of exceptional events that manage to reach the public domain.[33] As

32. Rasmussen, *The Risk Society at War*, 4.

33. For a discussion of the psychological reasons for this, see Cherie D. Maestas et al., "Shifting the Blame: Federalism, Media, and Public Assignment of Blame Following Hurricane Katrina," *Publius: The Journal of Federalism* 38, no. 4 (2008): 609–32.

individual attacks gain greater salience in directing public opinion, the pressures to manipulate information about these becomes stronger for the policymakers involved. Such efforts are also more likely to succeed because of the secrecy and limited oversight. With so much left unknown about the nature of cyberwarfare, and radical differences in competency in determining risk potentials, cyberweapons heighten the importance of perception. From the public's perspective, what is actually happening in the cyber domain is less politically important than what is perceived as happening. A country may be hemorrhaging classified information and have espionage software installed on every computer, yet this may be invisible to the public in a way that conventional military defeats would not be. Conversely, a honeypot defensive strategy of allowing attackers to penetrate computer systems could be highly effective while still undermining public confidence in information security if news of breaches is publicized.

Nowhere is the salience of perception management clearer than from the attribution problem. Virtually every major cyberattack is surrounded by controversy over who launched it. Identification is challenging because of the tendency to route attacks through third-party systems and to conceal or fake forensic evidence that could be used to identify a source. Even if attributional challenges can be solved, this usually takes a long time and a large investment of resources. "The forensics of attribution can rarely, if ever, give immediate results and can take days if not weeks to provide solid technical evidence."[34] Eun and Aßmann go so far as to say that "Determining the real aggressor is impossible unless the aggressor admits to it."[35] Doubt about the source of attacks and the long, complex investigative procedures offers attackers a degree of plausible deniability. Moreover, lack of attributional certainty heightens the risk of misattribution. Those reeling from an attack may be too quick to identify the wrong culprit, which could misdirect attention away from the real attacker and inspire hostility from those who are incorrectly blamed.

Already a pattern has emerged in the aftermath of cyberattacks as the victims blame familiar adversaries and recriminations are exchanged without much progress toward substantiating responsibility attributions or resolving disputes.[36] Again, the 2016 US presidential election is revealing. After years of investigation by intelligence agencies, the question of whether Russia influence the election remains hotly contested along party lines in the United

34. John B. Sheldon, "Geopolitics and Cyber Power: Why Geography Still Matters," *American Foreign Policy Interests* 36, no. 5 (2014): 288–89.

35. Eun and Aßmann, "Cyberwar," 13.

36. Marcus Schulzke, "The Politics of Attributing Blame for Cyberattacks and the Costs of Uncertainty," *Perspectives on Politics* 16, no. 4 (2018): 954–68; Kaplan, *Dark Territory*.

States. A Russian attack against the Olympics in 2018 showed that the country's efforts at concealing its activities are becoming more sophisticated over time, and in particular that it is becoming more skilled at planting false clues pointing to other possible culprits.[37] It is doubtful that any conventional attack could foster the same level of disagreement about responsibility.

Digital Battlefields

The physical infrastructure of cyberspace matters. The Internet does not exist in a vacuum outside space but rather arises from the servers, routers, and cables that store data and transmit it between machines. These systems can be mapped out geographically,[38] and there is strategic benefit in controlling the physical underpinnings of networks. The United States' control over the Internet's physical infrastructure puts it in a particularly strong position to intercept communications and monitor activities. There are places that lack Internet connectivity and those that are more vulnerable because they are more heavily integrated. Individual attacks may also have a physical component. Air-gapped systems that cannot be accessed remotely may have to be attacked by infiltrating the guarded facilities directly or introducing a piece of infected hardware. Stuxnet was delivered via infected USB drives, and follow-up attacks on the Natanz facility were likewise planned to involve compromised machinery. The physical security around computer systems is therefore a vital part of defense against cyberattacks.

Cyberwarfare has not obviated geographical considerations, but it has changed the nature of the battlespace, expanding it far beyond the scope of conventional military operations in at least three ways. First, cyberattacks allow rivals to strike each other even when they are not geographically proximate or competing for control over a particular territory. Conventional wars tend to work according to fairly clear geographical logics when it comes to the causes of war and the means of fighting.[39] Neighbors tend to fight more frequently because of disputes over borders, territories, and resources. The domains of combat are generally confined to areas that are mutually accessible and convenient. Wars between France and Germany, for example, are waged in those countries and their smaller neighbors, rather than being

---

37. Andy Greenberg, *Sandworm* (New York: Anchor, 2019).

38. Megan Garber, "What the Internet Actually Looks Like," *The Atlantic,* July 13, 2012, http://www.theatlantic.com/technology/archive/2012/07/what-the-internet-actually-looks-like/259815/

39. John Keegan, *A History of Warfare* (New York: Vintage, 1994).

fought in some distant and geographically unconnected domain. Cyberattacks do not require a proximate physical staging point from which to strike, and they are rarely carried out with geographical aims in mind (Russia's attacks on Georgia and Ukraine are noteworthy exceptions to this). This adds to attributional uncertainty and its associated diffusion of risk for the attacker, as an attack could potentially be launched by any state or nonstate actor, even if there is no physical link to the adversary.

Second, cyberwarfare helps to break the connection between territorial control and military strength. States are usually able to convert the former into the latter during conventional wars, such as by mobilizing their populations and extracting resources that can be used to produce war materiel. Geographical positioning has traditionally given states such powerful advantages or disadvantages that it has been a decisive factor in securing victory. For example, during the American Civil War, the North established an overwhelming geographical advantage by controlling the country's largest population centers, industrial regions, and ports. When it comes to cyberwarfare, physical resources are helpful, but they are far from decisive. Larger and wealthier countries find it easier to invest in cyber capacities, but these resources are not as directly convertible into coercive power as a large population or good strategic placement. Small states and nonstate actors can compete with much larger adversaries that would overwhelm them in conventional battles. Small actors will never be able to match the full scope of states' cyberwarfare capabilities, nor will they have the same capacities for integrating cyberattacks with physical attacks. Their cyberattacks will likewise probably not achieve anything like the scale of destruction wrought by conventional weapons, but, much like nomadic raids of the past and recent terrorist attacks, small cyberattacks can have a cumulative effect. At the very least, they divert state resources and achieve minor victories against isolated targets.

Third, the effects of cyberwarfare are less geographically confined than those of conventional strikes. Every networked system in the world is a possible source of attack, an accomplice, or a victim. A virus that is initially directed against a computer in a single country can spread to systems around the world. The ILOVEYOU virus was initially launched on May 5, 2000, and infected roughly forty-five million computers by the time it was contained.[40] In the case of ILOVEYOU, the wide distribution was built into the system because the virus e-mailed itself to everyone in a victim's address book. ILOVEYOU was deployed by criminals, and it is unlikely that any

40. Mark Ward, "A Decade on From the ILOVEYOU Bug," BBC, May 4, 2010, https://www.bbc.com/news/10095957

attacks coming from states, which is to say, attacks that may qualify as acts of cyberwarfare, would be intentionally indiscriminate because of the risks to the attacker's own computers or those of private individuals and corporations. However, the spread of a virus may be unintended. Once released, a virus takes on a life of its own, as in the case of Stuxnet, which was uncovered because it left the Natanz facility. States launching viruses even face the risk that any malicious code they release might ultimately come back to strike unprotected friendly systems. This tendency to expand geographically beyond a particular theater of war sets cyberattacks beyond the geographical confines upheld by the conventional warfare of the Westphalian system.

Selective Strikes

The risks of cyberattacks being deflected back onto the attacker, providing a road map for adversaries' own attacks, and spreading globally are powerful incentives to pursue greater individuation in attacks. Cyberwarfare is highly individuated in the sense that it is usually directed narrowly against a specific target, with success depending on limiting collateral damage to other systems. Indeed, one of the best ways of demarcating cyberattacks as a mode of warfare employed by states from cyberattacks as criminal activities done with malicious or financial goals in mind, is that the former are usually far more focused. Cyberattacks launched by states are best able to achieve the delicate balance of risk when they are carefully designed to avoid spreading too far or causing excessive damage. We can see the focus of attacks from the Stuxnet worm, which was designed to infect only a narrow range of computers with a specific configuration of centrifuges operated by Siemens PLCs. Stuxnet ultimately spread far beyond those machines, but it did not activate the payload in other nuclear control facilities and was powerless in other computers because the target had been so carefully selected. Stuxnet's relatives, Duqu and Flame, were narrower still. Flame was estimated to have infected around one thousand computers belonging primarily to government officials in the Middle East. It was also possible to transmit a kill command to the worms, at which point they would erase all traces of themselves from the targeted computers to avoid detection. Individuation therefore limited the prospects of these worms causing collateral damage or serving as blueprints for other exploits.

Attacking a narrow range of individuated targets facilitates risk management and perception management at the point of attack. The more specific the target and controlled the payload, the more likely an attack will go unnoticed by the victim. The longer an attack remains covert, the greater the prospects of success and evading the hazards associated with detection. This in turn facilitates perception management in a broader sense by either avoiding attention or at least ensuring enough distance from the attack to provide plausible deniability about a state's involvement in it. China has repeatedly entered US government and commercial networks to steal information, yet this has not provoked the same level of public outcry as malicious criminal attacks such as the ILOVEYOU worm or WannaCry, which spread rampantly through private computers to create a general sense of threat. In cases where the attack is quickly detected, individuation helps to limit the backlash. North Korea's attack on Sony Pictures in retaliation for making *The Interview*, a film showing a fictional assassination of Kim Jong Un, was an overt reprisal. However, it was narrowly focused on a specific company, so there was a low chance of triggering a strong response. If the United States had retaliated using any kind of conventional weaponry, it is likely that North Korea would have been seen as the victim of an unjust escalation of force.

Cyberattacks are not only individuated in the sense of the computer systems being struck but also in terms of the human victims. Usually only a few people have knowledge of the target systems, so only a few people have to be treated as defenders. The practice of social engineering involves manipulating a person into granting access to the target system, and therefore depends on correctly identifying the individuals involved and personal weaknesses that can be exploited. Famed computer hacker Kevin Mitnick describes impersonating dozens of different people to carry out his attacks and specifically targeting other members of organizations who could provide the vital information he needed to access protected systems.[41] He argues that hackers are apt to prey on vulnerable individuals because "the *human* factor is truly security's weakest link."[42]

This was vividly illustrated by the leak of thousands of Democratic National Committee e-mails, which came from Russian hackers targeting John Podesta while he was the chairman of the Clinton campaign. The attack was simple. Podesta received an e-mail purporting to be from Google and asking him to follow a link, then input security credentials. He forwarded the request to his IT department, which responded to say that it was "legitimate"; only later it turned out that the technician responsible had mistyped

41. Kevin D. Mitnick and William L. Simon, *The Ghost in the Wires: My Adventures as the World's Most Wanted Hacker* (New York: Little, Brown, 2011).

42. Kevin D. Mitnick and William L. Simon, *The Art of Deception* (Indianapolis: Wiley, 2002), 3.

"illegitimate."[43] With this reassurance Podesta entered his information, allowing hackers into his e-mail accounts and ultimately leading to the release of documents that irreversibly damaged Hillary Clinton's election campaign. It was a prime example of how much could be achieved by a fairly primitive cyberattack directed at key actors who failed to take adequate precautions.

Cyberattacks and Disruption of Norms

As I argued at the outset of the chapter, cyberwarfare is difficult to conceptualize within a conventional war framework. This likewise makes it difficult to say whether norms that were developed to restrict the conduct of conventional wars can be applied to digital hostilities. For one thing, cyberattacks are beyond the *jus ad bellum* criteria that are used to determine when war is justified. Just cause for war arises when a state is defending against an aggressive territorial encroachment or, by some accounts, when mounting a humanitarian intervention to protect people from atrocities. Cyberattacks may be aggressive, but they are not aggressive in either of these ways. The harm they inflict is not territorial in nature; attacks do not directly seize ground and are rarely used to facilitate this goal. Disruption is the more common objective. At the same time, because the targets are computer systems and not people, it is difficult to imagine a humanitarian basis for retaliating against cyberattacks that do not inflict casualties.

This raises the question of whether there can be just cause for defending against a cyberattack using conventional weaponry or whether the only permissible response is a retaliatory cyberattack. The challenge is made more arduous when the principle of last resort is added. This principle demands that belligerents pursue all peaceful alternatives that could reasonably be expected to resolve a dispute before resorting to violence. When it comes to cyberattacks, it is not always clear what response may be warranted or who has actually launched the attack. Even when responsibility seems fairly clear there are often some lingering doubts that prevent a definitive assignment of blame, leaving considerable uncertainty about where to direct diplomatic efforts, demonstrations of force, or sanctions.

An attack that destroys a country's infrastructure could arguably constitute grounds for a conventional war if the attack is severe enough to threaten

43. David Smith, "WikiLeaks Emails: What They Revealed about the Clinton Campaign's Mechanics," *The Guardian*, November 6, 2016, https://www.theguardian.com/us-news/2016/nov/06/wikileaks-emails-hillary-clinton-campaign-john-podesta

the lives of those living in the victim country, but at present it seems unlikely that cyberattacks will take this form in most cases. As I have emphasized, the appeal of cyberweapons lies in their utility as coercive tools that fall short of inflicting this kind of destruction. They are most advantageous when deployed in a way that forces the defender into the moral quandary of how, and whether, to retaliate within the confines of established Westphalian norms. That is to say, they are most advantageous when deployed in accordance with Sun Tzu's dictates, outside the confines of decisive battle. Those employing cyberattacks have strong incentives to keep their efforts small. Limited strikes can avoid provoking strong retaliation, or at least cause a defender that launches a disproportionately violent reprisal to concede the moral high ground. The normative problem that cyberwarfare raises when it comes to the justification of war is therefore central to what makes it so attractive. As with attributional uncertainty, this normative ambiguity becomes one of the attacker's greatest strengths by limiting the victims' ability to respond with anything besides its own cyberattacks or subtle diplomatic coercion.

With cyberwarfare straining core just war principles like just cause and last resort, it may be tempting to condemn it in the interest of preserving normative clarity. Cyberweapons could be declared inherently immoral (*mala in se*) because they fall outside the boundaries established by prevailing norms. Yet this conclusion is undesirable because cyberwarfare introduces so many potential benefits that should be attractive to anyone who subscribes to just war theory's goal of limiting the physical harm inflicted by war. As Rid points out, cyberattacks "are making previously violent interactions less violent. Only in the twenty-first century has it become possible for armed forces to cripple radar stations and missile launchers without having to bomb an adversary's air defense system and kill its personnel and possibly civilians in the process."[44] Cyberwarfare is normatively attractive as a way of achieving important political goals without incurring the high costs that are typically associated with violence, especially violence against civilians. We are therefore left with a paradox: cyberattacks interfere with compliance with some just war precepts, especially in terms of *jus ad bellum*, but they also advance the overall goal of restricting warfare and generally comply with *jus in bello* exhortations to use discriminate and proportionate force.

In addition to raising contrary normative costs and benefits, cyberwarfare makes it easier than ever to mobilize just war theory as a weapon. It is in

44. Rid, *Cyber War Will Not Take Place*, viii.

a sense a manifestation of the weaponization of international law in lawfare,[45] but taken more broadly to incorporate elements of *jus ad bellum* and *jus in bello* that have not been codified as law. By heightening compliance with certain elements of just war theory, especially proportionality and discrimination, cyberweapons give aggressors some claim to the moral high ground. Victims are then in the undesirable position of having to choose whether they should go beyond the just war framework by retaliating with force or whether to respond with reciprocal cyberattacks. In most cases, the incentive is to keep the exchanges confined to the cyber domain, lest the actors involved be perceived as deviating from the norms of war and escalating hostilities. Throughout this process there is apt to be a collective benefit from an overall diminution in the extent of violent conflicts as states are able to pursue their goals without resorting to conventional military force. However, the reduced likelihood of violent reprisal comes at the cost of more secrecy about the operations being conducted and more normative confusion.

The Rise of Hackers

Cyberwarfare joins drone warfare in harkening a new era in which the instrumentality of war triumphs over its existential meaning, with disruptive consequences for the military profession and its norms. Like drone strikes, cyberattacks are profoundly unromantic and unheroic according to traditional conceptions of martial virtue. They are launched by people sitting behind computers, working on lines of code without ever facing their enemies in open combat. There are, no doubt, moments of excitement when puzzles are solved, exploits are successful, or attacks are uncovered, but these are secret moments that cyberwarriors keep carefully hidden. Victory does not lead to ticker tape parades down Broadway or televised award ceremonies. There will be no monument to the creators of Stuxnet. There is certainly no popular narrative of cyberwarfare akin to John Wayne movies or first-person shooter video games, and given the nature of the work being done by the programmers, it seems unlikely that any will be created any time soon. The goal of cyberwarfare is not to win glory in pitched battles but to triumph quietly over an unknowing enemy; the goal is to win without the decisive confrontation that is the essence of war's existential appeal.

45. For discussions of lawfare, see David Kennedy, *Of War and Law* (Princeton: Princeton University Press, 2006); Orde F. Kittrie, *Lawfare: Law as a Weapon of War* (Oxford: Oxford University Press, 2016).

Cyberwarfare is transforming military institutions from inside and outside, forcing them to reconsider their membership and training while also challenging their relevance and potentially undermining their power. Internally, cyberwarfare is linked to the rise of technicians. More soldiers and intelligence agents perform computer-mediated tasks. They do work that previous generations of soldiers would not have recognized and that they might not have considered to be war at all. The mode of operations drives a deep wedge between the kinds of battlefield experiences that have unified soldiers across generations and their own mediated, computerized view of conflict. The Russian hackers who assisted the invasion of Georgia in 2008 doubtless had radically different experiences than soldiers on the ground, to the point that the two groups would have little basis for building an identity around a shared conception of military service.

James Cook defends the applicability of just war theory to cyberwarfare with an effort to show that code warriors are akin to drone operators and that armed forces are already recognizing their status as combatants. He argues that "those who labor across the interface of cyberspace are not working in any important sense differently than aviators or navigators."[46] "Certainly cyber operators might be targets of enemy attack, and one can imagine scenarios where they are especially endangered. But it seems more likely that many will never face the dangers that, say, infantrymen are used to. The cyber operators' daily grind might resemble that of drone pilots."[47] Cook also notes that the US Air Force took a radical step in 2005 when it declared cyberwarriors to be "operators" in the same sense as drone pilots and missile controllers, then later gave cyberwarfare a place in its mission statement alongside air and space operations.[48] This is an apt characterization of how the status of these digital warriors has grown in recent years, and it highlights the rate at which the practice of cyberwarfare is developing. To remain competitive, armed forces must incorporate cyberwarfare into their doctrines even as the associated norms and the professional obligations of cyberwarriors are deeply uncertain.

Cook is correct in thinking that there are already analogues for cyberwarriors in drone and missile operators. However, the problem is that all of these roles stand in sharp contrast to the face-to-face violence that is at the heart of the warrior ideal. The mediated activities he cites may resemble each other,

---

46. James Cook, "'Cyberation' and Just War Doctrine: A Response to Randall Dipert," *Journal of Military Ethics* 9, no. 4 (2010): 411–23, 414–15.

47. Cook, "'Cyberation' and Just War Doctrine," 419.

48. Cook, "'Cyberation' and Just War Doctrine," 419.

but they collectively constitute a radical departure from war as it was waged for virtually all of human history prior to the Cold War. They signal the same rise of instrumental war over existential war, and therefore the same challenge to traditional norms based around directly confronting the enemy. It is not just that cyberwarriors are coming into being as a new type of combatant but that they are also doing so alongside an array of similar types of new warriors who operate emerging weapons systems. My goal in looking at multiple topics throughout this book is to emphasize that the challenge to the military profession is pervasive, beyond any single domain or technology. The breadth of this challenge means that change is overdetermined in the sense that efforts made to limit the impact of one or two innovations would not stem the tide.

Many of the technicians producing, executing, or defending against cyberattacks are not members of the military at all but rather belong to various intelligence services. Even describing these personnel as "members" of intelligence services may be a step too far when so many are private contractors who work with government agencies without being formally inducted. The five leading intelligence collection contractors in the United States have around ninety thousand employees who hold security clearances.[49] The Russian government has repeatedly sought assistance from private supporters during its campaigns.[50] It has likewise evaded responsibility by blaming private actors for initiating cyberattacks on its behalf. This is a point I return to in chapter 4, but for now it is important to note that intelligence personnel and contractors are outside the scope of how the military profession has traditionally been defined, thereby blurring the profession's boundaries.

Another challenge for the military profession, and to the idea of conventional war, is that civilians are tentatively pulled into the world of cyberwarfare. Civilians are rarely expected to bear arms in defense against conventional threats and would compromise their status as civilians by doing so, yet when it comes to cyberwarfare civilians are expected to be active participants in defense. They are thereby transported into cyberwarfare along its periphery. Virtually all civilian computer users are expected to participate in the fighting to some extent, often with small acts of personal security, such as changing passwords regularly, staying alert to phishing attempts, and updat-

49. Tim Shorrock, "5 Corporations Now Dominate Our Privatized Intelligence Industry," *The Nation*, September 8, 2016, https://www.thenation.com/article/five-corporations-now-dominate-our-privatized-intelligence-industry/

50. Maurer, *Cyber Mercenaries.*

ing their software. These may seem to be trivial security procedures, especially since they have become such a routine background condition of life, but these constitute a kind of self-defense against cyberattacks. These are the steps needed to protect government systems from attack and to prevent personal computers from being compromised and integrated into botnets. Civilians who fail to exercise adequate precautions may even become participants in attacks in a very real sense of being part of the physical infrastructure that makes attacks possible. After all, negligence may cause civilians to inadvertently release sensitive information, introduce a worm hidden on a flash drive into sensitive systems, allow a computer to become part of the next DDoS attack, or inadvertently e-mail a virus to others.

States retain a significant advantage over nonstate actors when it comes to their resources and the numbers of personnel they can employ. Nevertheless, cyberwarfare opens competition between states and nonstate actors far more than conventional warfare, and it is possible to find instances of people who would ordinarily be classified as civilians creating severe security disruptions. When pursuing nonpolitical motives, hackers are better classified as criminals than as enemy combatants, but when the extent of their attacks can equal those arising from states to pose serious threats to national security, it is difficult to draw a sharp line between criminal and political violence. For example, the hacktivists of Anonymous launched dozens of attacks to achieve political objectives, and it is at times described as a terrorist organization, which brings it into the gray area between law enforcement and war normative paradigms.[51] Hackers also feed into interstate conflicts by developing new exploits that can be copied by states or sold to them directly. Studies of the market for zero day exploits indicate that states are among hackers' main customers.[52] Given the continually adversarial and collaborative interactions between states and nonstate actors in the cyber domain, it is impossible to establish a sharp conventional war division between the two.

Conclusion

The conventional wars of the nineteenth and twentieth centuries were won by belligerents who could mobilize massive armies and the most destructive lethal technologies. They were epitomized by the face-off between power-

51. Olson, *We Are Anonymous.*
52. Menn, "SPECIAL REPORT."

ful conventional militaries, with both sides sending as many rounds and bodies at the other as possible to achieve decisive victories. Strategy and tactics mattered, of course. Operational doctrines of maneuver helped to achieve overwhelming victories. However, the most important innovations came in the management of new technologies and associated doctrines, such as those for armored, air, and submarine warfare. Even the great operational feats of the past century—the German blitzkrieg, Israel's triumph in the Six Day War, the liberation of Kuwait—could only be achieved with the assistance of new technologies of conventional warfare. The rise of cyberwarfare in the closing years of the twentieth century and into the twenty-first holds the promise of broadening conflict in a new direction. It privileges deception and nonlethal attacks above overwhelming force, and this in turn privileges a different political sensibility. It rewards those who are able to use cyberattacks to make subtle and possibly even imperceptible shifts in the political landscape, rather than those who would seek to produce mass physical destruction. It likewise rewards those who can manipulate perceptions and use the fear of attack as a weapon in its own right.

States and nonstate actors alike benefit from being able to pursue their goals without appearing to do so publicly and while having plausible deniability if they are detected. When narrowly focused, cyberattacks are highly individuated and largely cohere with the most important normative constraints on war, thereby managing the risk of striking the wrong target and while giving perpetrators considerable control over how they are perceived. Doing this entails a major shift in how wars are waged, one that is so extreme that it becomes difficult to reach agreement about whether digital aggression is a type of warfare at all. Such attacks fall largely outside the scope of traditional military roles and bypass the geocentric orientation of conventional warfare. And yet these attacks are carried out by political actors and with the same kinds of instrumental goals that would have previously belonged to conventional military forces. It therefore represents a new domain of warfare based on a unique form of mediation that upsets the geographical, normative, and professional characteristics of armed conflict.

3 • War without Casualties

Nonlethal Weapons and Military Policing

Conventional war is typically characterized as an effort to kill enemy combatants, and, as the previous chapter discussed, this is one of the characteristics by which cyberattacks ostensibly fall short of the definition of war. However, the use of nonlethal violence throughout the history of warfare shows that it is misleading to think that killing is an essential feature of war beyond this conventional understanding. It is imperative to avoid mistaking the style of war at a particular moment, in this case, during the Westphalian era, with a timeless ideal that all wars must satisfy. This would lead the definition of war to be contingent on ephemeral trends in how violence is enacted in a particular epoch. Some tribal societies staged battles more as a ritualized system of dispute reconciliation than an effort to kill, even discontinuing battle if anyone died in the fray.[1] Where war is based on a bilateral agreement to disable opponents rather than kill them, it tends to be a ceremonial affair that is nonlethal by mutual agreement.[2] It proceeds according to relatively clear, though probably unwritten, rules to ensure that the contest never escalates into the more violent confrontation that it always threatens to become. Nonlethal violence can also be pursued asymmetrically, without assuming mutually binding rules. Aztec warriors attempted to subdue their opponents with clubs instead of killing them.[3] Amassing prisoners was a sign of a warrior's courage, especially when adversaries were armed with more lethal implements. Colonial operations to capture slaves likewise privilege capture over killing and rely heavily on nonlethal tools.

1. John Keegan, *A History of Warfare* (New York: Vintage, 1994).
2. Keegan, *A History of Warfare*, 131.
3. Ross Hassig, *Aztec Warfare: Imperial Expansion and Political Control* (Norman: University of Oklahoma Press, 1988).

Contemporary nonlethal weapons employed in military contexts partake of both traditions of nonlethal combat, the asymmetric and the symmetric/ceremonial. Nonlethal weapons are generally deployed in asymmetric conflicts and used to subdue adversaries who are not following the norms of conventional war, but with the goal of imposing strict normative order rather than capturing sacrificial victims or slaves, which is more characteristic of symmetrical/ceremonial nonlethal violence. Whereas bilateral uses of nonlethal weapons are often grounded in ritual and shared norms that unite opposing belligerents, the contemporary emergence of nonlethal weapons shows the appeal of fighting with minimal force because the norms are in jeopardy. Nonlethal weapons are meant to improve armed forces' abilities for fighting in accordance with the principles of discrimination and proportionality at a time when doing this is harder than ever because of the prevalence of opponents (and allies) who blend in with civilian populations. Nonlethal weapons reflect the new condition of unconventional fighting in which soldiers are rarely sure who is an enemy combatant and the misuse of lethal violence threatens to destroy the fragile normative order. The appearance of nonlethal weapons on contemporary battlefields reveals an attempt to remake the practice of war, transforming it from killing between groups to a kind of law enforcement activity that leaves those on both sides relatively safe.

Unlike premodern nonlethal weapons, contemporary variants are informed by an instrumental rather than an existential logic. That is to say, they are the product of political calculations, not an attempt to give warriors more opportunities for displaying their prowess in the face of danger. Three instrumental goals are especially important, and they contrast sharply with the conventional war goal of inflicting maximum destruction on the enemy. First, bringing conflict into populated areas without the risk of injuring or killing innocent people. This helps to transform war from an attritional contest between opposing armed forces into an effort to seize control of population centers, and especially symbolically important capital cities. Second, minimizing visible, and *especially countable*, losses on all sides of the conflict (including among the ranks of enemy combatants). Nonlethal weapons may inflict pain and suffering, but usually in ways outside observers cannot see or measure. This also serves the broader mission of asserting absolute control over the bodies of people in contested areas to the extent that there is power over life as well as death. Third, giving soldiers weapons they can use against innocent people with minimal fear of seriously injuring or killing them. Nonlethal weapons reflect a tacit acknowledgment that force will sometimes be misapplied against civilians, and an effort to reassert centralized political control over operations to prevent such incidents from undermining the legitimacy of the overall mission.

As with the other weapons I have discussed thus far, nonlethal weapons comprise a large class of many different technologies. These include devices that disable people or machines with sounds, smells, electrical currents, blunt force, or physical restraints. The label "nonlethal" is controversial. "The basic counter has been that no weapon is free from the risk of fatalities."[4] This is an important point to bear in mind when seeking clarity about this broad category. Even the most restrained nonlethal weapons can produce lethal effects depending on how the weapons are employed. For example, a device that disables vehicles without interacting at all with the human body could still cause fatal collisions. At the same time, it is essential to avoid watering down this category too much by suggesting that a weapon's lethality is entirely dependent on what the attacker means to do with it. Nonlethal weapons can injure and kill, and they can be used with this intent, but they remain a special category because the weapons themselves embody an intent to minimize violence under most circumstances.

Wars without Death

Weapons are embedded in sociopolitical contexts, which means that "the question as to why, at certain times and places, some weapons should be more acceptable than others depends on factors other than the characteristics of those weapons themselves. In fact, its causes are to be found deep within a civilization and the attitudes that it takes to military conflict."[5] This is borne out by nonlethal weapons, the development and implementation of which are heavily shaped by disillusionment with conventional interstate war as a method of resolving political disputes and the need for coercive tools that can be used outside the conventional war framework.

Western warfare is often described as a brutal affair in which warriors proved their worth by killing and enduring the threat of being killed in turn.[6] This characterization is somewhat misleading, as polities have routinely sought to avoid participating in war that is designed to maximize lethality.

4. Brian Rappert, *Nonlethal Weapons as Legitimizing Forces? Technology, Politics, and the Management of Conflict* (London: Frank Cass, 2003), 18.

5. Martin Van Creveld, *Technology and War: From 2000 B.C. to the Present* (New York: Macmillan, 1991), 72.

6. Victor Davis Hanson, *The Western Way of War: Infantry Battle in Classical Greece* (New York: Random House, 2013).

Reciprocal exposure to death may be edifying for warriors, but it is profoundly risky when pursuing political objectives. Throughout the Middle Ages, it was common for European belligerents to avoid decisive confrontations that would result in high casualties. They preferred capturing supplies, towns, and key fortifications.[7] Pitched battles were rare and lasted just a few hours. Captured soldiers were frequently ransomed or paroled, which is a sign that battle was partly governed by mutual agreement to minimize its intensity. Even the horrific Hundred Years War was largely conducted through raids and sieges.[8] There were, in other words, efforts to minimize killing long before the advent of modern nonlethal weapons. This kind of limited warfare was sporadically interrupted by deliberately destructive conflicts, especially when religious differences were the reason for fighting, but it would be misleading to say that an intent to inflict casualties was an inherent part of pre-Westphalian war.

The Westphalian era was characterized by a quest for shorter and more decisive clashes meant to settle disputes quickly by inflicting maximum damage. Weigley argues that "this strategy of annihilation of the enemy through battle was a rational response to the difficulties of achieving the objects of policy through war."[9] Conventional warfare thus changed lethality from an optional aspect of war into a defining feature. The rapid escalation of destructive power throughout the Westphalian era, and especially during the industrialization of war in the nineteenth and twentieth centuries, was ultimately self-defeating. Rather than shortening wars and reducing the overall costs of solving political disagreements on the battlefield, overpowered munitions increased the costs of violence to such an extent that interstate wars could have apocalyptic consequences. Nonlethal weapons constitute a partial step back to pre-Westphalian war. Although states have not abandoned their conventional weapons, and probably will not do so in the foreseeable future, the necessities of resolving political disputes with lower degrees of lethality have compelled them to search for more usable weapons.

Like earlier changes in the conduct of war, the addition of nonlethal weapons to military arsenals reveals broad social and political trends. The tendency to promote military values as prerequisites for good citizenship and war as an existential quest for national glory, which prevailed during the height of the Westphalian system from the Napoleonic Wars to the Second

7. Jeremy Black, *Warfare in the Eighteenth Century* (London: Cassell, 2002); Brian Bond, *The Pursuit of Victory* (New York: Oxford University Press, 1998).

8. Jonathan Sumption, *The Hundred Years War, Vol. 1: Trial by Battle* (Philadelphia: University of Pennsylvania Press, 1999).

9. Weigley, *The Age of Battles*, 536.

World War, has fallen out of favor. Instead of rallying their populations behind great military crusades, states now attempt to fight in ways that are minimally disruptive and largely invisible to the general public. This is true even in the relatively rare cases of conventional warfare, such as the Falklands War and the Gulf War. Nonlethal weapons address the demand for separating war from everyday life by reducing the likelihood of the overt physical harm that is the most vivid visual reminder that wars are taking place. To the extent that states are able to fight using nonlethal weapons, rather than conventional weapons, they can assert that the fighting does not rise to the level that would qualify as war.

The need for nonlethal weapons has been strongest in unconventional settings in which priorities have shifted from defeating enemy armed forces to pacifying foreign civilian populations. Unconventional conflicts are also the settings in which armed forces are best able to minimize violence without forfeiting their ability to successfully coerce their adversaries. Here nonlethal weapons can be used directly against noncompliant civilians and against armed nonuniformed combatants. Soldiers must be prepared to confront a host of situations in which lethal force is usually disproportionate: stopping vehicles at checkpoints, dispersing crowds, searching civilians' homes, and escorting convoys through urban areas amid civilian traffic. The popularity of population-centric counterinsurgency and counterterrorism missions has had an especially strong influence on the technology.

Among the oldest nonlethal weapons still in service are chemical irritants like CN, CS, and CR gas. CN was first used during World War I to compel enemy soldiers to fight while wearing bulky chemical warfare equipment. This shows that nonlethal gases can be used in conventional operations. However, innovation took the logical path toward more lethal chemical agents when the destruction of enemy armies was the main objective. Nonlethal gases found a new home in counterinsurgency operations. CS and CR were employed in the 1950s by British colonial forces, then in the 1960s by the US military in Vietnam. They became a staple of asymmetric conflicts in Northern Ireland, South Africa, the Philippines, Israel, and elsewhere.[10] Nonlethal weapons as a general class have largely lost their conventional war orientation. Rather than being developed for conventional uses then redirected into other applications, the most innovative emerging nonlethal technologies are usually designed specifically for law enforcement and unconventional warfare. This signals that armed forces

10. Anna Feigenbaum, *Tear Gas: From the Battlefields of WWI to the Streets of Today* (New York: Verso Books, 2017).

are not simply repurposing conventional warfare tools when they are involved in unconventional operations but rather are developing substantial capacities purpose-built for the latter.

Covering the myriad nonlethal technologies would require an entire book, but a few noteworthy examples can illustrate how they function. Rubber and plastic bullets were introduced by the British military in the 1970s and achieved prominence during operations in Northern Ireland.[11] Israel regularly employs these "kinetic impact rounds" and refines them by testing new materials. The active denial system (ADS) that was created by the US military directs electromagnetic energy to produce an intense burning sensation that rarely causes lasting harm.[12] The weapon is designed to be used in crowd control situations and has the advantage of being effective against large groups of people without diminished strength or much risk of inadvertently killing someone. Stun guns cause temporary neuromuscular incapacitation. Acoustic weapons are designed to incapacitate victims with nausea, pain, or even the disruption of biological processes.[13] Other nonlethal weapons are less futuristic. Among the devices under development to halt vehicles and people are nets that would physically seize them.

Koplow emphasizes that the extent to which nonlethal weapons are discriminate would depend on which nonlethal weapons are used, and he notes the marked differences between the various instruments that fall into this category. This makes it essential to remember that any generalizations about nonlethal weapons may have exceptions when it comes to certain devices. There are also important differences in the nature of the harms nonlethal weapons inflict even when these fall short of killing. Gross argues that some nonlethal devices might be abhorrent and possibly even worse than conventional weapons if they inflict permanent disability, such as if a laser were used to blind enemies without killing them.[14] This judgment coincides with the United Nations Convention on Certain Conventional Weapons Protocol IV, which prohibits blinding lasers.[15] The use of torture throughout the War

---

11. Feigenbaum, *Tear Gas*.

12. For more inform about the ADS, see Robert M. McNab and Richard L. Scott, "Nonlethal Weapons and the Long Tail of Warfare," *Small Wars & Insurgencies* 20, no. 1 (2009): 141–59.

13. Jürgen Altmann, "Acoustic Weapons: A Prospective Assessment," *Science & Global Security* 9 (2001): 165–234.

14. Michael L. Gross, *Moral Dilemmas of Modern War: Torture, Assassination, and Blackmail in an Age of Asymmetric Conflict* (New York: Cambridge University Press, 2009), 65.

15. https://treaties.un.org/doc/Treaties/1995/10/19951013%2001-30%20AM/Ch_XXVI_02_ap.pdf

on Terror has raised additional concerns of nonlethal weapons being used against prisoners to inflict extreme pain without causing death. "The problem with targeting NLWs solely as a product used for torture is that, 'torture is an issue of human intent, not devices.' Any object can be used as an instrument of torture or ill-treatment."[16] There is, in other words, a problem of being unable to link form and function that makes these highly adaptable weapons amenable to a range of different uses that not only span policing and wartime roles but that also differ sharply in their justifiability.

Fighting by the Rules

Recent enthusiasm for using nonlethal weapons reflects a normative crisis brought about by the ways in which state armed forces designed for conventional warfare have struggled to adjust to unconventional warfare. States fight unconventional adversaries who exploit ambiguities in international law and just war theory, especially by blending in with indigenous civilian populations. The principle of discrimination, which is one of the most basic and important elements of just war theory and international law, forbids belligerents from targeting civilians or recklessly endangering them. It also arguably requires belligerents to exercise heightened care whenever their actions could inadvertently cause civilian casualties.[17] The promise of protection gives civilians an incentive to distance themselves from the war effort, thereby limiting the number of people who may be attacked. These two mechanisms should in principle help to ensure that by reducing the number of participants, wars are more proportionate than they would otherwise be.

Armed forces intending to abide by the principle of discrimination face a growing array of barriers that inhibit compliance. Unconventional opponents may avoid wearing uniforms, employ human shields, mix with the civilian population, occupy civilian structures, and use civilian vehicles. A central argument from those who favor deploying nonlethal weapons is that these allow soldiers to carefully calibrate their responses to the threats at hand.[18] Soldiers could be more liberal in their uses of force for self-defense, but at a lower overall cost to civilians who would be injured rather than

16. McNab and Scott, "Nonlethal Weapons and the Long Tail of Warfare."

17. See the discussion of due care in Michael Walzer, *Arguing About War* (New Haven: Yale University Press, 2004).

18. McNab and Scott, "Nonlethal Weapons and the Long Tail of Warfare."

killed. For example, during the Battle of Mogadishu, American soldiers were confronted by large mixed groups of armed and unarmed Somalis.[19] They initially held their fire, but as the battle developed, they shot directly into the crowds, later saying that their right to self-defense made this response justifiable. Armed with nonlethal weapons, they might have been able to disperse the crowds without producing as many civilian casualties.

The potential benefits of nonlethal weapons are clear from how they could alleviate escalation-of-force decisions. These arise when soldiers operating in a potentially hostile area, such as in a city where insurgents are active, have to evaluate potential threats and respond appropriately. When they are armed with conventional weapons, soldiers' usually face escalation-of-force decisions as a stark choice about whether to employ lethal force against unknown, nonuniformed persons who may or may not be threatening. For example, soldiers at a checkpoint may have to decide whether an approaching person is a civilian or an enemy fighter preparing to launch an attack. The person in question may appear to be a civilian, but possibly concealing a weapon or a suicide bomb beneath their clothes.[20] Darkness, fog, dust, and other environmental conditions further impede efforts to discern whether a person is a civilian or an insurgent emulating one. Vehicles pose even more serious challenges. Some conceal heavily armed insurgents and vehicle-borne explosives. Others are packed full of civilians. Soldiers searching buildings, conducting patrols, guarding traffic control points, and escorting convoys continually face escalation-of-force decisions while contending with fear, uncertainty, sleep deprivation, and other cognitive impediments.

Militaries have developed escalation-of-force procedures that are meant to alleviate the stark decision of whether to use lethal force under conditions of uncertainty by deterring the target before the fateful choice must be made. Warning signs, verbal warnings, and warning shots are the most common methods of deterrence. However, because deterrents are often ineffective or misunderstood, soldiers' escalation-of-force decisions typically fall back into the binary of "take a shot or take a chance [of being killed]."[21] One US Army

---

19. Mark Bowden, *Black Hawk Down: A Story of Modern War* (New York: Atlantic Monthly Press, 1999).

20. Marcus Schulzke, "Ethically Insoluble Dilemmas in War," *Journal of Military Ethics* 12, no. 2 (2013): 95–110; Marcus Schulzke, "The Unintended Consequences of War: Self-Defense and Violence against Civilians in Ground Combat Operations," *International Studies Perspectives* (forthcoming); Marcus Schulzke, *Pursuing Moral Warfare: Ethical Theory and Practice in Counterinsurgency Operations* (Washington, DC: Georgetown University Press, 2019).

21. Dexter Filkins, "A Nation at War: In the Field Marines; Either Take a Shot or Take a Chance." *New York Times*, March 29, 2003.

smartcard that was designed to help soldiers decide when to attack oncoming vehicles at checkpoints offers an escalation-of-force procedure soldiers are supposed to employ in what the card estimates to be a 11.6-second window of opportunity from when a vehicle is first sighted to when it reaches the traffic control point and could potentially detonate a suicide bomb.[22] That gives soldiers less than twelve seconds to attempt deterrence and decide whether to use lethal force. When these kinds of decisions play out repeatedly every day in contested areas, the predictable result is that lethal violence is frequently misdirected against civilians. It is possible that attacks against civilians under these circumstances could be avoided if soldiers use nonlethal weapons whenever they are not actively engaged in combat. For example, unidentified persons who fail to comply with verbal instructions at a checkpoint could be shot with a stun gun rather than with a rifle.

Nonlethal weapons may also help to solve the problem of how to deal with civilians who pose low-level threats or who facilitate insurgent attacks. Palestinians rely heavily on the tactic of throwing nonlethal projectiles like rocks and bricks in their confrontations with Israeli forces.[23] This caught the Israel Defense Forces off guard during the First Intifada, when soldiers who were trained for conventional warfare were involved in street clashes with rioters. It resulted in some tragic escalation-of-force decisions, with Israeli soldiers firing live ammunition into crowds or shooting extremities to cripple the protestors. British and American soldiers faced similar challenges in Northern Ireland, Afghanistan, and Iraq, among other places. They also routinely encountered civilians who directed mortar attacks or triggered improvised explosive devices using phones.[24] Nonlethal weapons address these problems through an individuation of violence that comes from more precisely calibrating levels of force to the adversary. Threatening civilians could be disabled rather than being killed. Moreover, the mere capacity to launch nonlethal attacks could help to establish a more credible deterrent against low-level violence from civilians in the first place.

Nonlethal weapons reconfigure the binary choice of "kill or do not kill" as a spectrum that includes different levels of force between these options. At the very least, nonlethal weapons create an intermediary point between the two extremes; they may add more complexity if the nonlethal weapons come in varying degrees of strength. Thus nonlethal weapons "enable the possessor to target a particular person, place, or thing with greater precision and to project

22. Center for Army Lessons Learned, *Traffic Control Point Operations (GTA 90-01-005)* (2010).

23. Sergio Catignani, *Israeli Counter-Insurgency and the Intifadas: Dilemmas of a Conventional Army* (New York: Routledge, 2008).

24. Schulzke, *Pursuing Moral Warfare.*

a hostile effect in a more discrete, temporary, circumscribed manner."[25] Efforts to make more fine-grained use-of-force decisions require fundamentally rethinking the status of civilians and the principle of discrimination. Rather than a clear black and white admonition to not attack civilians, soldiers armed with nonlethal weapons could face a more delicate task of determining when and how civilians may be attacked in that intermediate zone between the extremes. This increase of options would grant armed forces greater flexibility in situations when civilians are acting hostilely or in which they cannot be easily differentiated from combatants, and it would likely lower overall civilian casualties. However, it would also increase the cognitive burden on soldiers tasked with acting across a broad spectrum of policing and war-fighting roles. Soldiers would have to make highly individuated decisions about how to respond on a person-by-person basis, rather than simply categorizing people in terms of the civilian/combatant binary.

One could argue that use-of-force decisions are adequately covered by existing norms of war, which prohibit the targeting of civilians. For example, armed forces could reserve nonlethal weapons for use against armed enemy combatants and prohibit their use against civilians. However, this response has two problems. First, it puts more weight on the question of whether certain types of military operations actually count as war. As I have argued, any line between war and peace that could provide normative clarity is ambiguous and becoming more difficult to locate, particularly in the types of unconventional conflicts in which nonlethal weapons are most useful. If enemy fighters are considered criminals, then the norms of war do not provide the correct framework, and soldiers would need to think more like police officers. The record of conflicts at this point in the century suggests that we cannot assume such conceptual certainty and that most conflicts will tend to be on the margins, where norms relating to the treatment of civilians are unclear. Second, strictly confining nonlethal weapons to existing use-of-force guidelines would forfeit their advantages. Much of the appeal of nonlethal weapons lies in the prospect of reducing the costs of attacks that are misdirected against civilians or that are intentionally directed against mildly threatening civilians. It seems unlikely that armed forces developing nonlethal weapons would undercut their value by requiring that they be used with the same restraint as lethal weapons. This is especially true if nonlethal weapons are not sufficiently powerful to deter armed enemies.

Nonlethal weapons display an admirable desire to minimize harm to

---

25. David A. Koplow, *Death by Moderation: The U.S. Military's Quest for Useable Weapons* (New York: Cambridge University Press, 2010), x.

civilians who are at risk of being inadvertently attacked and to reduce the level of force used against weakly armed enemies. However, these weapons pose their own moral challenges. The potential gain in civilian protection depends on assuming that soldiers would be freer to attack civilians who are mildly threatening or who cannot be differentiated from enemy combatants. That is to say, it depends on interpreting the principle of discrimination narrowly as forbidding lethal violence against civilians, but permitting the more liberal use of force that is apt to follow from soldiers being allowed to protect themselves with nonlethal means. Some commentators respond to this challenge by treating it as grounds for rejecting nonlethal weapons. Pauline Kaurin argues that "As soon as we entertain the permissibility of targeting some non-combatants, we have stepped over a crucial moral line. The principle of discrimination relates not just to killing non-combatants, but to intentionally targeting non-combatants."[26] She cites the attempt to rescue hostages during the standoff at Moscow's Dubrovka Theater in 2002 as an example of this. Russian security forces indiscriminately dispersed gas into the theater to disable everyone inside, with the result that 133 hostages were killed along with the terrorists. Russian officials even refused to identify the gas used, which interfered with efforts to provide effective medical treatment for the victims. Kaurin makes a good point, but it is difficult to see how a lack of nonlethal weapons could have led to a better outcome for the hostages. Perhaps more would have been killed if security forces had stormed the building. Moreover, it is difficult to generalize based on one case involving gas when so many other types of nonlethal weapons are available. The real problem could be the type of nonlethal weapon selected or how it was used.

Both sides in the debate over the permissibility of nonlethal weapons are justified in thinking that the opposing position puts civilians in jeopardy. This is because nonlethal weapons sublimate the principle of noncombatant immunity in the Hegelian sense: they preserve it, cancel it, and supersede it simultaneously.[27] These weapons preserve civilian immunity insofar as they manifest an unprecedented desire to reduce the extent of civilian suffering inflicted by wars. They cancel civilian immunity by sanctioning what the principle of discrimination is supposed to prohibit—the targeting of civilians—by making it easier for soldiers to attack first and investigate later. Finally, nonlethal weapons supersede civilian immunity because the combi-

---

26. Pauline Kaurin, "With Fear and Trembling: An Ethical Framework for Nonlethal Weapons," *Journal of Military Ethics* 9, no. 1 (2010): 100–114, 105.

27. For a more detailed discussion of this concept see Walter Kaufmann, *Nietzsche: Philosopher, Psychologist, Antichrist* (Princeton: Princeton University Press, 1974), 237.

nation of preserving it and canceling can only be reconciled by reaching a new understanding of civilian immunity, according to which civilians may be targeted with nonlethal force if they are engaged in low-intensity violence or if soldiers have reasonable suspicions that an unknown person is a combatant who would, in the absence of nonlethal options, be subjected to a lethal attack. This sublimation makes it impossible to give a simple answer about whether nonlethal weapons are simply good or bad when it comes to civilian protection. A claim one way or the other would overstate the level of moral clarity when nonlethal weapons are apt to reduce the number of civilians who are killed while still being an affront to the concept of civilian immunity, possibly increasing the number of nonlethal injuries, and still leaving the prospect of some civilians being inadvertently killed. We are left with the unenviable choice between upholding a strict conception of civilian immunity, at the cost of continued lethal attacks against civilians, or privileging lives saved over principles, at the cost of taking a more permissive attitude when applying the principle of discrimination.

Invisible Violence

Twenty-first-century armed forces face unprecedented demands of being accountable to public opinion and fighting in ways that civilians at home and watching from third-party countries will support, or at least tolerate. That is, perception management has risen to prominence as a vital strategic demand that armed forces must take into account. The US Marine field manual *Military Operations in Urbanized Terrain* warns that "commanders at all levels must understand the impact that media representation will have on the accomplishment of operational and strategic objectives."[28] The US military's *Joint Urban Operations* manual likewise says that "any individual observing military operations with access to the Internet can post information and images that have the potential to reach millions of people, influencing their perceptions about military operations. This is particularly true in urban environments where operations are often conducted under the observation of the citizenry."[29]

Nonlethal weapons provide a vital tool in the quest to manage perceptions more effectively. Their creation signals that a military recognizes the

---

28. U.S. Marine Corps, *US Marine Field Manual on Military Operations in Urbanized Terrain, MCWP 3–35.3* (Washington, DC: Department of the Navy, 1998), 1–21.

29. Joint Chiefs of Staff, *Joint Urban Operations* (Washington, DC, 2013), I-7.

importance of minimizing civilian casualties and maintaining high ethical standards. The image of armed forces researching and testing nonlethal technologies may be a useful public relations move in its own right, regardless of whether these are used much in practice. It is probably no accident that some of the most futuristic devices regularly appear in YouTube videos, military documentaries, and videogames long before they enter military service. The mere pursuit of these weapons is used to signal the desire to prevent civilian casualties and reduce the overall levels of suffering wars inflict.

Nonlethal weapons can also facilitate perception management when they are used. Older nonlethal devices, such as gases and rubber bullets, can cause serious physical injuries that are easily observable. However, more advanced nonlethal weapons are typically designed to inflict less serious and less visible harm, often only causing pain when the weapon is aimed directly at the adversary. This is the case for weapons like the active denial system (ADS) and long range acoustic device (LRAD). The objectively perceivable physical harm of lethal weapons and earlier nonlethal weapons is replaced by a more subjective perceptual experience of harm that cannot be accessed by third-party observers. The harms then become incommensurable. This means that the associated risks cannot solely be gauged by focusing narrowly on the immediate victims of attack. They must be assessed more broadly by considering the systemic impact of lowering perceived harm to civilians even as attacks against them persist.

Postmodernist critics of the Gulf War claimed that it marked a turn into the hyperreal, with war becoming more simulation than an actual event.[30] Jean Baudrillard went so far as to say that "the Gulf War did not take place."[31] The point was not about whether the event really happened but rather about how it was perceived from a distance, as a war without the fighting and dying that is associated with war. From the perspective of audiences around the world, the signs of war were erased via physical mediation. Nonlethal weapons are taking this perceptual shift one step further. Whereas critique of the Gulf War was based on how the casualties were borne by one side and then hidden from the public's view by selective media coverage, nonlethal weapons may make it possible for wars to be conducted with a dramatically reduced level of visible violence even without efforts to deceive or frame conflicts in self-serving ways. If wounds are invisible, war ceases to be a visual spectacle of suffering.

30. Paul Virilio, *Desert Screen: War at the Speed of Light* (New York: Continuum, 2002).

31. Jean Baudrillard, *The Gulf War Did Not Take Place* (Bloomington: Indiana University Press, 1995).

Of particular importance is that this perceptual shift could upset the restrictive effects of casualty aversion. By many accounts, civilians have lost their enthusiasm for war. They resist fighting when there is a high likelihood of casualties, or lose confidence in ongoing wars as casualty numbers rise.[32] The true extent of this phenomenon is heavily debated, yet critics of new weapons systems (especially the three I discuss in this book) contend that it is an essential mechanism for imposing restrictions on when wars are waged. As Kaurin says, applying this reasoning to nonlethal weapons specifically, "War does and should cost. NLW should make war more humane, but never costless. Even if it were possible to make it costless, it would be morally problematic to do so."[33] Although I sympathize with Kaurin's concerns, I contend that this argument goes too far. The deterrence value of lethal violence is contentious. It is unclear how effective this type of deterrence is, the circumstances in which it takes root, and whether it is even real. It would be misguided to oppose technologies that hold a high probability of reducing the magnitude of suffering in the short term to gamble on a much less certain reduction in the incidence of war over the long term, at least until we have far more evidence about the comparative costs of these options.

Koplow presents a more cautious and plausible version of this argument, saying that "The illusion that NLW could enable us to fight sanitized, bloodless wars with minimal casualties on all sides may obscure some of the inherent costs and dangers of any realistic military engagement."[34] It seems unlikely that nonlethal weapons could induce this degree of overconfidence as they are currently used, since they tend to be relegated to specific roles and are rarely cited as a reason for being able to initiate wars more easily. However, this is a reasonable concern as capacities for nonlethal violence continue to expand, especially as militaries invest more heavily in counterinsurgency and humanitarian operations that may change their armaments and images to be less closely associated with traditional combat activities. From this perspective, these weapons have a manipulative character insofar as the perception management may be used for the nefarious purpose of masking the harms associated with war and making it easier for states to employ coercion.

32. Martin Shaw, *The New Western Way of War: Risk-Transfer War and Its Crisis in Iraq* (Malden, MA: Polity Press, 2005), 35.

33. Kaurin, "With Fear and Trembling," 27.

34. Koplow, *Death by Moderation*, 205.

Fantasies of Violence and Expectations of Passivity

War not only holds an existential meaning for soldiers but also for civilians. The romantic narratives of war often go unrealized for the countless soldiers who actually experience combat firsthand, those who have the romanticism shattered by real experiences of fighting. However, for those of us who do not share in the trauma of combat, the romantic images can persist without the same kind of harsh contact with reality. Such images are supported in countless ways, from films about the heroes of past wars, to sporting events celebrating veterans, to video games that allow players to imaginatively project themselves into combat as "virtual citizen-soldiers."[35] Popular culture engagements with war strive to maintain the sense that war is in some sense a force for good and that, no matter how violent, it may help us realize our highest values as individuals and as nations. Above all, the romanticization of war depends on stories about individual sacrifice, courage, and loyalty in the face of danger. It is no accident that popular media about war focus on the most dangerous military occupations (e.g., infantry combat or covert missions, rather than maintenance or supply requisitioning), as it depends on an enduring narrative of war as an existential experience in which soldiers find themselves and develop fraternal links to their comrades by enduring danger and loss.

There is a paradox embedded in this enthusiasm for violence. It is often said that the civilian public of Western countries is highly sensitive to casualties, such that support for war deteriorates quickly as casualties are sustained. Those who subscribe to this view generally see the United States as being especially vulnerable to casualty aversion, with the Vietnam War, intervention in Somalia, and interventions in the Balkans during the 1990s serving as prime examples of how quickly civilian support may evaporate. William Martel contends that "In Vietnam, mounting U.S. casualties contributed to the collapse of public support for the war because it fostered the perception that victory was not a realistic outcome and that intervention was not worth the cost in lives and resources."[36] John Mueller has traced a decline in support for wars as casualties increased during the US wars in Korea, Vietnam, Somalia, and Iraq.[37] Edward Luttwak argues that "military force collides with the

35. Roger Stahl, *Militainment, Inc.: War, Media, and Popular Culture* (New York: Routledge, 2009).

36. William C. Martel, *Victory in War: Foundations of Modern Strategy* (Cambridge: Cambridge University Press, 2011), 163.

37. John E. Mueller, *War, Presidents and Public Opinion* (New York: Wiley, 1973); John Mueller, *Policy and Opinion in the Gulf War* (Chicago: University of Chicago Press, 1994).

general refusal of the American public to sanction interventions in place after place without end."[38] This poses a potential risk to national security, suggesting that the will to fight can disappear even during strategically important wars. Depending on how it is framed, casualty aversion may apply to allied military personnel who are lost, or to foreign civilians, or both. At its root, the phenomenon is about a desire for war to either be avoided or to be remade as a virtually cost-free enterprise.

Other commentators question this narrative and contend that public support for wars persists despite the casualties incurred. Charles Hyde says that this view is a myth,[39] and Eric Larson advises that "it is more accurate to say that the public hopes for low-to-no casualty operations but fears a very different outcome."[40] Burk likewise says that "patterns of public support for peacekeeping missions reveal no irrational or knee-jerk reactions based on a putative unwillingness to tolerate casualties."[41] However, regardless of whether minimizing casualties is essential to maintaining public support, it is evident that many politicians and military officers *perceive* this pressure and attempt to avoid incurring any drops in domestic or international support they may suffer from losing their own soldiers or inflicting foreign civilian casualties.[42] Adopting nonlethal weapons presents them with a tradeoff between minimizing the prospective costs of war in an effort to manage perceptions by way of managing risks or permitting some violence in the interest of sustaining some semblance of war as a violent enterprise, complete with all the allure that goes along with that. They may not encounter this problem in such clear terms, with a stark decision between these alternatives, yet decisions relating to the use of nonlethal weapons are essentially this. They revolve around the challenge of managing risks in such a way that will strike the delicate balance needed to maintain support.

The paradox, then, is that maintaining war's attractiveness depends on celebrating danger and heroic death, but successfully prosecuting war is a matter of minimizing risk and avoiding casualties that could trigger real or imagined casualty aversion. There is no way of truly having it both ways, of

38. Edward N. Luttwak, "Toward Post-Heroic Warfare," *Foreign Affairs* (May/June 1995).

39. Charles K. Hyde, *Casualty Aversion: Implications for Policy Makers and Senior Military Officers* (Newport, RI: Naval War College, 2000).

40. Eric V. Larson, *Casualties and Consensus: The Historical Role of Casualties in Domestic Support for U.S. Military Operations* (Santa Monica, CA: RAND, 1996), 101.

41. James Burk, "Public Support for Peacekeeping in Lebanon and Somalia: Assessing the Casualties Hypothesis," *Political Science Quarterly* 114, no. 1 (1999): 53–78, 77.

42. Hyde, *Casualty Aversion*.

waging risky wars with minimal casualties. This makes it essential to manage perceptions to give the appearance that both goals are achieved simultaneously. Nonlethal weapons are ideally suited for this. Lower levels of violence can help to reduce civilian casualties and expand soldiers' opportunities for using force in self-defense; at the same time soldiers can participate in the more dramatic and heroic ground combat operations with opportunities for heroism.

Countering Nonlethal Weapons

Historically, the central goal for armed forces was destroying the enemy and capturing territory, while risks were primarily understood as being linked to the enemy's reciprocal attacks. Weapons progressed through advancements in lethality, distance, and armor, which allowed militaries to destroy enemies more effectively while also limiting the prospects of being attacked in turn. Histories of military innovation trace a progression of escalating lethality over time, culminating in the enormously destructive wars of the twentieth century.[43] And even now it is common for military strength to be measured in terms of total destructive power. Lethal coercive force is the clearest metric of states' relative strength, but one that all too often fails to have adequate predictive power because of the decline of attritional warfare.

Nonlethal weapons mark a fundamentally different approach to risk management than the one that has prevailed in previous conflicts. Rather than risk management hinging on offensive destructive power and defensive force protection, it is now characterized by a much broader effort to minimize harm that extends to civilians, and even to enemy fighters. "For the first time, we are directing our defense laboratory wizardry along the azimuth of self-restraint, and it is important to tease out the reasons why this odd sort of revolution is occurring and what its implications may be."[44] Many weapons participate in the developmental shift toward usability, including the drones and cyberweapons that I discussed in previous chapters, as well as precision guided munitions and electronic interference devices. This is a systematic trend that reflects a new style of strategic thinking that is not reducible to a

43. Stephen Peter Rosen, *Winning the Next War: Innovation and the Modern Military* (Ithaca: Cornell University Press, 1991); Wayne E. Lee, *Waging War: Conflict, Culture, and Innovation in World History* (New York: Oxford University Press, 2016).

44. Koplow, *Death by Moderation*, 18.

single technology. However, nonlethal weapons are distinctive even among these usable weapons because they do not attempt to focus killing power narrowly but to actually bypass the need for lethality altogether.

Once again, the risk-management strategies fall into the familiar pattern of averting some risks while introducing new ones. Altmann's research provides excellent insight into the illusory effectiveness of many nonlethal weapons that are currently in development.[45] He argues that many of the claims made by weapons manufacturers are false and self-serving, as weapons cannot possibly perform as well as promised. Looking at acoustic weapons, he shows that these may only be able to disable people in controlled settings, and that even then they may be neutralized by earplugs or other inexpensive defenses. He finds that to be effective, acoustic weapons would have to be made far more powerful, to the extent that they would lose any claim to being more humane weapons by causing severe pain and permanent hearing loss.

Research on other nonlethal weapons likewise suggests that in general the more a weapon's effects are toned down in the interest of preventing harm, the easier it may be for enemies to circumvent the weapons.[46] Weapons are not simply lethal or nonlethal; they span a range of different force options. Some nonlethal weapons, such as the gas used by Russian security forces during the theater hostage rescue, come very close to qualifying as lethal weapons because the narrow safety margins make them extremely difficult to employ. On the other hand, nonlethal devices like nets and sound waves are at the far extreme of devices that are unlikely to produce fatalities, but easier to counteract. Nonlethal weapons will fall into the same back and forth between weapons and countermeasures that we see with other weapon systems, but they may be even more susceptible to being circumvented by relatively low-tech defenses.

Promises of reduced violence are not as straightforwardly beneficial to leaders as they initially seem. Sapolsky and Shapiro express concern that nonlethal weapons may actually create a trap for armed forces that face the demand of fighting in ways that coincide with these high expectations. They argue that "the use of nonlethal alternatives may carry special penalties in terms of unrealistic expectations that a battle with nonlethal weapons will cause zero deaths."[47] They go on to say that even if nonlethal weapons do work as promised, the resulting injuries might be distasteful to a public that has been primed to expect restraint. Weapons that blind or cause disfiguring

45. Altmann, "Acoustic Weapons."
46. Rappert, *Nonlethal Weapons as Legitimising Forces?*
47. Sapolsky and Shapiro, "Casualties, Technology, and America's Future Wars," 122.

injuries could even be more upsetting than those that simply kill. Thus, Sapolsky and Shapiro conclude that "There are political as well as military minefields to cross in the quest to reduce killing." Research on drone strikes suggests that this evaluation is correct, by showing that members of the public are more tolerant of foreign civilian casualties when these are accurately forecast than when they are inflicted after misleading promises of zero casualties.[48] In other words, even when casualties are the same, public opinion responds more harshly to attacks that involve broken promises of low or no casualties. When it comes to drones, the promises of lowered lethality may or may not be borne out in practice depending on how they are used, but these machines are at least expected to kill enemies. For nonlethal weapons, the promise of de-escalation is so absolute that it is present in the name itself. It is a promise that is destined to be violated no matter how assiduously nonlethal weapons are employed, as even the strictest uses are apt to result in some accidental deaths.

Soldiers on the Beat

The goals of minimizing force that nonlethal weapons are designed to satisfy are not intrinsic to the military profession or the warrior ethos. Rather, they are external demands that are imposed from outside by political leaders attempting to manage conflicts and civilian audiences who may react harshly to superfluous violence. Doctrinal statements tend to mask this tension to preserve a clear continuum between political and military goals, yet the disjuncture routinely appears when soldiers speak openly about the profession and its norms. A recurrent complaint among military personnel fighting in recent counterinsurgency operations is that they wanted to be soldiers, not police officers. They report feeling that the role is degraded when they must conduct searches, guard traffic control points, and carry out investigations to find their enemies.[49] Many soldiers join the military because they want to experience combat.[50] Popular culture feeds these promises by focusing on combat as the defining experience of the military profession. Training, peacetime, and the boredom of waiting for a fight are brief asides, with the overall focus being pitched battles. Already soldiers

48. James Igoe Walsh, "Precision Weapons, Civilian Casualties, and Support for the Use of Force," *Political Psychology* 36, no. 5 (2015): 507–23.
49. Schulzke, *Pursuing Moral Warfare.*
50. For examples of soldiers professing a desire for combat see Bowden, *Black Hawk Down*, 9.

comment on how brief combat is compared to what they had envisioned and how much time they spend waiting for battles that may never come. This can cause considerable resentment from the soldiers who dedicate their lives to the preparation for war.

In his memoir *Jarhead*, US Marine Anthony Swofford laments his failure to truly experience combat as he envisioned it, which deprives him of the status so many soldiers aspire to. "To be a marine, a true marine, you must kill. With all of your training, all of your expertise, if you don't kill you're not a combatant." He goes on to say that "Sometimes you wish you'd killed an Iraqi soldier. . . . You consider yourself less of a marine and even less of a man for not having killed while at combat."[51] Such is the appeal of killing that it can provoke these feelings of desire and self-doubt, and this experience will be denied to many soldiers who are sent to war armed with nonlethal weapons. This may ultimately save countless soldiers from the more serious trauma of mistakenly killing innocent people, but it is a change that violates their expectations of war and their sense of what it means to be a soldier.

Nonlethal weapons are symptomatic of role strain that is building up to an identity crisis. Studies of nonlethal technologies tend to speak about police officers and soldiers in the same breath, drawing parallels between their operations and advocating shared procedures.[52] Nonlethal weapons tend to flow easily between wartime and law enforcement applications, especially when counterinsurgency resembles a kind of transnational policing. The movement of weapons from one domain into the other is both a symptom and cause of blurring boundaries between law enforcement and war. It is a symptom because the appeal of nonlethal weapons in both domains signals a deep similarity in the actions performed. It is a cause because this sharing provides opportunities for cross-training, technology sharing, and collaboration. David Kilcullen argues that soldiers in future operations will need to behave far more like police officers, even to the point that they substantially revise certain aspects of the military culture.[53]

A profound change is underway as new sensibilities urge for greater restraint among military personnel. It lays the foundation for the sharing of technologies and values between military and law enforcement roles. Never-

51. Anthony Swofford, *Jarhead: A Marine's Chronicle of the Gulf War and Other Battles* (New York: Scribner, 2003), 247.

52. John B. Alexander, *Future War: Nonlethal Weapons in Twenty-First-Century Warfare* (New York: St. Martin's Press, 1999).

53. David Kilcullen, *Out of the Mountains: The Coming Age of the Urban Guerrilla* (New York: Oxford University Press, 2013).

theless, military training remains primarily focused on conventional war, as does the professional ethos.[54] The work of soldiers has not been shifted from war to policing but rather expanded to encompass both roles. Such broad training is inherently difficult because it aims to cover so much. It becomes all the more challenging when the different types of work belong to two fundamentally different security paradigms with their own norms, practices, and political logics. The inevitable result is that one framework or the other will win out. Thus far, it has meant that soldiers continue to train for war and to be encultured into institutions that are designed for war while also being instructed in low-intensity operations and the use of nonlethal weapons. The world's leading militaries continue to see themselves first and foremost as war-fighting institutions, with counterinsurgency often being treated as an unwelcome distraction that they have to adapt to, to the detriment of conventional war-fighting capabilities.[55] Adopting nonlethal weapons would present a serious material threat to this sensibility by arming soldiers with tools that are generally unsuited for conventional war and that signal a strong commitment to engaging in law-enforcement activities.

Convergence of Spaces

The battlefield is not only constituted by the physical terrain that combatants occupy when fighting but also by the distinctive norms and practices that constitute it as a separate sphere of life that can be marked with territorial boundaries. The battlefield is delineated by an imaginary border that has profound political and normative importance but that is incredibly delicate. The line is, after all, a social convention put in place by mutual agreement between belligerents that can be dissolved when that agreement disappears. Contemporary military discourse reflects the dissolution of this line, with language shifting from talk of "battlefields" to a more totalizing conception of the "battlespace." As Stephen Graham notes, the concept of battlespace points to geographically unlimited war.

---

54. Paul Robinson, Nigel De Lee, and Don Carrick, eds., *Ethics Education in the Military* (Hampshire: Ashgate, 2008); John Wesley Masland and Laurence I. Radway, *Soldiers and Scholars: Military Education and National Policy* (Princeton: Princeton University Press, 2015).

55. Thomas R. Mockaitis, *The COIN Conundrum: The Future of Counterinsurgency and US Land Power* (Carlisle: US Army War College–Strategic Studies Institute, 2016).

> Nothing lies outside the multidimensional and multiscale concept of battlespace, temporally or geographically. Battlespace has no front and no back, no start and no end. . . . The concept of battlespace thus encompasses everything from the molecular scales of genetic engineering and nanotechnology, through the everyday sites, spaces and experiences of city life, to the planetary spheres of inner and outer space or the internet's globe-straddling "cyberspace."[56]

For Graham, this makes it possible to expand security practices into everyday life, but the converse is also true, and just as significant. Security practices of domestic life expand into war. This is exactly what we see with nonlethal weapons, and it is no accident that armed forces have begun investing in these more heavily in recent decades, at the same time as the concept of the battlespace has taken on greater prominence as a framing concept for strategic thinking.

The introduction of nonlethal weapons into the battlespace alongside the introduction of military security practices into domestic spaces marks a change in how geography relates to conflict. For one thing, this contributes to the general trend of military actions and law enforcement becoming more similar. Nonlethal weapons make spaces of conflict look less like battlefields and more like a domestic setting in which violence should be governed by law enforcement norms. As I have argued, more widespread reliance on nonlethal weapons is sure to aggravate efforts to distinguish between these domains, their respective norms, and the roles that are grounded in them. This is evidence that some of the key physical manifestations of war that assist in distinguishing spaces of battle from spaces of peace are under threat from nonlethal weapons. When wars and law enforcement are enacted using similar methods, it is more difficult to point to substantive differences as constituting the geographical boundary between battlespace and domestic spaces.

Building on this, there is a deeper implication about the nature of sovereignty. Implicit in totalizing claims about the battlespace and efforts to conduct security operations of war and peace in the same manner is a tacit claim about sovereignty expanding beyond a state's borders into any spaces where its military forces are actively using nonlethal weapons and framing their operations along the lines of the battlespace concept. Nonlethal weapons are not so much weapons of war as they are tools of governance by which states

56. Stephen Graham, "When Life Itself Is War: On the Urbanization of Military and Security Doctrine," *International Journal of Urban and Regional Research* 36, no. 1 (2012): 136–55, 138.

assert military power where normative constraints would not permit them to do this according to conventional conceptions of war. They are symptoms of powerful armed forces acting as global police forces. This is not necessarily bad. After all, this type of policing is what informs humanitarian interventions aimed at protecting people from atrocities. The decline of the Westphalian system through global policing opens a future in which human security can become a shared responsibility and more plausibly guaranteed against malicious governments. This trend is apt to carry mixed effects, with interventionism helping to save lives while also creating a mechanism for asserting control over foreign countries from behind a mask of beneficence.

The Stuff of War

Mediation is an important piece of the puzzle of how nonlethal weapons are changing the geography of conflict. The physical mediation of violence during war is usually concerned with offensive and defensive lethal force. As I pointed out in the introduction, war is virtually always a mediated activity to some degree. Even those who fight with swords or spears, or bullets and bombs, employ forms of mediation that set them apart from their targets. Sometimes the distance is only a few feet and sometimes it is miles, yet the goal is the same: to kill enemies more effectively and at longer ranges to prevent retaliation. Armor is also a form of mediation, as it places a material between the attacker's weapon and the user. Again the basic function is the same whether it is worn on the body, set in a vehicle, or abstracted into a missile defense shield. These forms of offensive and defensive physical mediation and the competition between them are central to the development of war over time, and in virtually all instances they have to do with the management of lethal force, either the ability to deliver it by creating risk imbalances that work against the receiver or the ability to defend against it by creating risk imbalances that work against the attacker.

Nonlethal weapons are defined by their alternative forms of mediation in the sense that they alter the physical substance of fighting. Offensively they replace the steel, lead, and explosive materials characteristic of lethal weapons with a broad range of materials and energy systems that were not previously used in wars. Bullets become plastic or rubber to reduce their penetration. Controlled energy beams cause burning sensations without searing skin. Bursts of electricity are used to stun and disable. Some nonlethal weapons even involve the creation of new materials. Special foams have been made to cover targets and reduce their mobility. New chemical compounds are devised to irritate skin, produce noxious odors, or impede vision, without inflicting serious injury. Nonlethal weapons development is a search for different types of mediation than those employed in conventional weapons. On a more fundamental level this marks the emergence of a new material basis for war, one that effectively militarizes an array of substances that previously had no business in fighting. Sound, energy, air waves, gas, and exotic new substances become instruments of war and of population control in equal measure.

And mediation goes beyond the stuff projected out of nonlethal weapons when they also violate basic norms of bodily integrity. To kill someone in combat may deprive the victim of life, but it does so while still allowing the victim a certain measure of dignity that follows from being able to fight and die as a person in control of their own faculties. Older nonlethal weapons, such as rubber or plastic bullets and gas, resemble lethal implements of war in this respect, as they are essentially weakened versions of the weapons that kill. By contrast, many nonlethal weapons of the twenty-first century use the victim's body as their medium of attack. The body becomes a liability, a source of weakness that is exploited by the weapons system. It is disabled and rendered useless by pulses of energy that prevent muscles from working. It is not destroyed by physical intrusion but instead becomes dysfunctional as its parts are turned on each other. Water molecules stored in cells are brought to excruciating temperatures to create intense burning sensations. Pain receptors are overwhelmed by invisible forces. There is a certain indignity to the method of attack. It enlists the victim's body as a collaborator that can stop working, either partly or wholly, rather than destroying it. Much as cyberweapons transform the defender's infrastructure into the weapon itself, nonlethal weapons mediate war through the bodies of those on the receiving end. Their bodies are not simply destroyed but controlled, sometimes down to a molecular level.

To appreciate the significance of bodily governance it is useful to draw on some insights from studies of torture. The parallels between nonlethal attacks and torture are fairly clear. In each case the goal is to exercise coercion through the body itself without allowing the subject to die. The same devices can even be deployed in both contexts, as nonlethal weapons can inflict significant pain during torture without producing much physical evidence. As Lauren Wilcox points out, "Bodies are tortured, but they are not allowed to die. Death operates as a limit on the torturer as well as on the agency of the prisoner. The prisoner cannot be killed because he must be made to speak."[57]

57. Lauren B. Wilcox, *Bodies of Violence: Theorizing Embodied Subjects in International Relations* (Oxford: Oxford University Press, 2015), 59.

To prevent death, or even visible signs of injury, is to exercise total power over the body, the power of a sovereign over its subjects. When belligerents employ lethal violence they hold the power to kill, but not to control their adversaries. By contrast, nonlethal violence extends biopolitical power by guaranteeing such total bodily control that killing becomes superfluous.

The use of nonlethal weapons even makes it possible to deny enemies opportunities to die when they seek to die. A glorious death in combat can be a profound symbolic claim about the deceased's dedication to a cause and the inherent worth of that cause. This type of death is one of the ideals of the existential conception of war. Many combatants seek to become martyrs for precisely this reason, sometimes even offering themselves up for suicide attacks in which the aim is to die in an act of devotion.[58] Wilcox contends that prisoners and especially victims of torture may attempt to reassert sovereignty over their bodies through hunger strikes and other acts of disobedience, but that even this is denied when states force-feed prisoners and refuse to let them die. Nonlethal weapons may likewise be a way of denying enemies the opportunity to die. Just as a nonlethal device may disable a civilian who could be mistaken for a suicide bomber, it may disable the bomber and frustrate the attempted suicide. Denying enemies the possibility of death, especially the kind of symbolic death that is so important for the existential conception of war, is not merely a frustration of an individual's purpose. It subverts a mode of warfare that has prevailed for centuries.[59] This is no small achievement given the importance that suicide attacks have taken on in the asymmetric wars in which states are most inclined to employ nonlethal weapons.

Conclusion

In this chapter I have sought to point out some of the paradoxical implications of nonlethal weapons in military roles. Nonlethal weapons make it possible to use more finely calibrated levels of force that can be directed at particular targets while minimizing the chances of accidentally killing someone. They have the potential to lower the costs of mistakes in escalation-of-force decisions. Unidentified people who appear to be threatening may be disabled, rather than forcing soldiers into the dilemma of whether or not to kill them. Nonlethal weapons preserve civilian immunity

58. Wilcox, *Bodies of Violence*, 59.

59. Here I am referring not just to suicide bombing but to any forlorn attacks inspired by the glory of martyrdom.

by making it possible to reduce the number of civilians who are inadvertently killed, and they threaten civilian immunity by making it easier to launch attacks directly against civilians. These contradictory processes converge to push for a transcendence of those norms and the emergence of new sensibilities that link warfare more closely to law enforcement. This does not mean that nonlethal weapons are objectionable. In many ways they are advantageous. Even when misused, they are apt to inflict fewer casualties than conventional weapons. The decline in the overall extent of the violence can therefore be considered an important advancement. My point is only that every advancement comes at a price. In this case, the price is a disruption of the security paradigms that further undermines an increasingly tenuous distinction between them. Nonlethal weapons in military contexts are best seen as introducing contrary advantages and disadvantages that are irrevocably linked and that must be taken together.

Nonlethal weapons are clearly distinct from drones and cyberweapons, and yet they show a continuity of underlying themes that unite the defining weapons of twenty-first-century military operations, especially when it comes to the weapons' political import. Nonlethal weapons are a vital tool for states waging asymmetric wars in which they must use force while still managing the perceptions of audiences at home, in the contested area, and in third-party countries. Producing them signals an admirable commitment to reducing the destructiveness of wars and protecting civilian welfare. At the same time these weapons mark an expansion of governmental power as bodies are turned against themselves and even denied the ability to die in combat. Alongside drones and cyberweapons, nonlethal weapons display the extension of state power into new domains that have only recently become potential sites of war. In particular, nonlethal weapons introduce new opportunities for asserting an ostensibly benign sovereignty over territories and bodies.

Part II • Organizational Innovations

4 • Buying Victory

Private Military Contractors and the Decline of Popular Warfare

Private military contractors exist when the demand for military capacities cannot be satisfied by conventional armed forces alone. The problem of demand outstripping supply affects weak and strong states alike, although usually for different reasons. For weak states, contractors bring professionalism. They serve as disinterested fighters or provide training to strengthen national militaries. These are vital capacities when local forces are corrupt, uncooperative, or ineffective. Of course, the moral and political implications of propping up weak states can vary considerably depending on what type of government is in place and whether the leadership is legitimate. Factions within weak states may also subvert the national government by creating private militias, as in the Philippines.[1] For states with strong centralized governments, contractors bring a diversity of skills and greater numbers. Contractors are sometimes needed to maintain logistical networks without diluting the military's strength by assigning soldiers to support work. They likewise serve as bodyguards, aid workers, prison guards, and civil servants, jobs that few regular soldiers are trained for. The growing prominence of contractors in the twenty-first century has much to do with the rise of unconventional wars in which even the most powerful militaries have faced challenges beyond the scope of the operations they were built for, such as policing and development assistance.

Contractors supplement conventional military strength by providing a greater range of capabilities and additional manpower. However, they are also profoundly threatening to states and their armed forces. Of all the cases I consider in this book, contractors pose the most direct challenge to the Westpha-

1. "Private Armies in the Philippines: Guns and Goons," *The Economist*, January 9–15, 2010, 45.

lian system because their mere existence undermines the state monopoly on coercive force. This challenge is most visible in developing countries where private militias achieve pre-eminence over local government militaries, but it is just as real for strong states dependent on private support. States gain some capacities for risk management by outsourcing security and logistical tasks, but averted risks are inevitably replaced by new ones. Because contractors can change loyalties, reliance on them opens up new risks associated with defection, poor performance, and public relations debacles. For example, contractors were useful in American and British efforts to maintain the perception of limited commitments in Afghanistan and Iraq during the War on Terror, but they raised their own image management problems after widely publicized attacks on civilians.[2] Moreover, each individual state that participates in the market undercuts the collective normative barriers against private involvement in wars and fuels the industry, opening a panoply of future risks from private actors. States that employ contractors pursue short-term objectives by creating a long-term threat to sovereign authority in matters of war.

Like the other innovations I discuss, contractors are not simply good or bad. Private military contractors help to end wars, protect diplomats, and are involved in humanitarian relief operations around the world. However, they also escalate hostilities through human rights abuses, introduce dangerous market forces into military affairs, and weaken accountability for use-of-force decisions. Understanding their contribution to conflict in the twenty-first century should encourage us to take stock of the assorted costs and benefits and to situate these in the broader context of thematic changes in the conduct of war.

Contractors come in many forms. I follow Singer in categorizing them into three types: those that fight, those that provide advice and training, and those that perform support roles.[3] My focus is on contractors who fight, though each type is important for characterizing the contractor phenomenon as a whole. This is particularly true because major companies offer multiple types of services. Contractors who participate in combat have the most profound implications for the future of armed conflict because they show the expansion of military force to cover a broader range of functions and present a threat to prevailing conceptions of military professionalism. Although support and training contractors are less overtly disruptive and

2. Jeremy Scahill, *Blackwater: The Rise of the World's Most Powerful Mercenary Army* (New York: Nation Books, 2007).

3. P. W. Singer, *Corporate Warriors: The Rise of the Privatized Military Industry* (Ithaca: Cornell University Press, 2008).

receive less attention in this chapter, they too signal a change insofar as they illustrate the continued growth of military support structures and the expansion of operations beyond the scope of what existing militaries are able to achieve. When these three types of contractors are taken together, they show a broad expansion of private enterprise into roles that have for more than a century fallen under the remit of state military forces.

Mercenarism is a perennial feature of war. Greek hoplites were hired by the Persian Empire. Sir John Hawkwood's White Company and other mercenary armies spent decades fighting the countless wars of Renaissance Italy. And Hessians fought alongside the British to retain control over its colonies in North America.[4] Before the twentieth century, armies found it impossible to consistently feed themselves, making access to markets essential. It was only with the dramatic expansion of militaries in size throughout the nineteenth century and the growth of transportation networks (especially railroads) that logistics fell under state control.[5] The popularity of mercenaries throughout much of human history, including rampant use of mercenaries during the Thirty Years War that was ended by the Treaty of Westphalia, had much to do with the fragility of sovereign authority. In most cases, polities could not afford to build and maintain their own armed forces, especially during peacetime. They also struggled to manage professional combatants, which were prone to criminality and rebelliousness, in domestic spaces where law enforcement capacities were likewise limited. Mercenaries were a cost-saving expedient that also forestalled the potential adverse consequences of arming certain segments of the population.

Westphalian states were largely defined by a consolidation of coercive power, which made it possible to fund and effectively manage standing armed forces. Westphalian concepts of sovereignty, according to which states hold a monopoly on the legitimate use of armed forces within their borders, were fundamentally opposed to contractors. And the Westphalian ideal of conventional war depended on creating norms of fighting that were best sustained by professional, state-controlled officer corps. In cases when combatant mercenaries survived into the era of state consolidation, it was usually because of close alignments with a particular patron state or because the mercenaries operated on the fringes of sovereign authority, where there was less capacity for maintaining armies and fewer dangers of them causing political instability. States almost

4. Alan Axelrod, *Mercenaries: A Guide to Private Armies and Private Military Companies* (Washington, DC: CQ Press, 2014).

5. Martin Van Creveld, *Supplying War: Logistics from Wallenstein to Patton* (Cambridge: Cambridge University Press, 2004).

always replaced private forces when they had the capacity to do so, and they did the same once the tools of industrialization allowed them to bring logistical support networks under centralized control.[6]

Military history shows a close link between the popularity of mercenaries and the nature of sovereign authority. The new mercenaries emerging in the twenty-first century illustrate the same point. For weak states, reliance on contractors is a sign that the Westphalian system was never truly complete and that, paradoxically, the vestiges of statehood can sometimes only be maintained by assenting to an alternative conception of the state as a pre-eminent, rather than monopolistic, user of force within a territory. For more powerful states, the change reflects the fragility of military power and the transformation of war into a persistent, global law enforcement activity that exceeds existing capacities. In both cases, mercenaries are novel in the sense that they are reviving a military organizational practice after several centuries of decline and giving it a distinctive new form shaped by contemporary political circumstances.

Before moving on it is important to clarify some terminology. I generally use the term "mercenary" when referring to private military forces that existed before the 1990s, and the term "contractor" when referring to those of the post–Cold War era. This is an artificial temporal distinction. However, it reflects an important change in private forces since the end of the Cold War: a rebranding effort to make private forces a legitimate part of state arsenals after centuries of being maligned. This name change is among the most distinctive efforts to manage perceptions about the use of privatized force, and one that betrays the signs of a market system in which branding is essential to building sales. This is also an instance in which ontogeny recapitulates phylogeny. Just as "mercenary" has morphed into "contractor," individual contractor firms continually change names, marketing strategies, and management to cultivate a good image.

Mercenaries, Past and Present

There are deep affinities between military organizational structures and technological innovations. Technologies set some of the key structural con-

6. The American Civil War provides excellent examples of how governments attempted to manage state-run and private logistical networks, and the advantages of privileging the former when this was possible. See Earl J. Hess, *Civil War Logistics* (Baton Rouge: Louisiana State University Press, 2017).

ditions that permit organizational structures to exist and function. Conversely, organizational structures chart the path of innovation and shape the implementation of new technologies. This relationship between technology and organization is evident from the history of mercenaries. Privatized military forces have historically been most attractive to employers when they performed specialized roles that required professionalization or when they offered esoteric technical expertise. As in civilian industries, the mastery of a technique or technology fostered a comparative advantage over competitors, to the point that many mercenaries became associated with specific weapons or tactics that provided a unique advantage. Early crossbowmen, halberdiers, and handgunners sold their services to European kings hoping to use revolutionary weapons to tip the balance of power on the battlefield. Going back earlier, the Greeks and Romans sought Rhodians as specialists in firing catapults.[7] And Persian emperors employed Greek mercenaries as heavy infantry. States continue to rely heavily on private assistance when venturing into new domains of conflict, as evidenced by the reliance on private actors to supply cyberwarfare capacities.[8]

Not all mercenaries have adopted innovative weapons. Some marketed themselves with tactical skills that had to be cultivated through consistent practice that was beyond the grasp of rank-and-file soldiers serving for short periods. Norse Vikings became the Varangian Guard protecting Byzantine emperors. German Landsknechts were experts in countering cavalry. Japanese ninjas were seen as "anti-samurai" because they were usually peasants who rejected the code of Bushido to launch covert operations.[9] This trend of tactical specialization continues as contractors pursue niche skills that lie outside those of regular militaries, such as protecting diplomats or guarding prisoners. Again, this illustrates private actors' penchant for finding gaps in the market where comparative advantage can yield the greatest benefits.

The historical stigmatization of mercenaries owes much to the social and political threat they posed. Even before the Westphalian era, their specialized weapons and tactics frequently deviated from normative restrictions designed to protect the status quo. For example, Genoese mercenaries employed crossbows even though those weapons had been banned by the

7. On these early approaches, see in particular John France, ed., *Mercenaries and Paid Men: The Mercenary Identity in the Middle Ages*. Proceedings of a Conference Held at University of Wales, Swansea, July 7–9, 2005, vol. 47 (Leiden: Brill, 2008); and Kelly DeVries and Robert Douglas Smith, *Medieval Military Technology* (Toronto: University of Toronto Press, 2012), 119.

8. Tim Maurer, *Cyber Mercenaries: The State, Hackers, and Power* (Cambridge: Cambridge University Press, 2018).

9. Axelrod, *Mercenaries*.

Second Lateran Council in 1139. Many special weapons and tactics undermined entrenched political interests. Mercenaries were countermeasures against knights, samurai, and other elites whose social standing depended on the monopolization of martial skill. They were likewise threatening to more inclusive republican governments. When deployed alongside or against citizens' militias, mercenaries upset the identity of civic participation in combat. Mercenaries were therefore often revolutionary figures whose influence reached far beyond the battlefield. Their military abilities had political effects. Whenever they threatened elites, they raised the prospect of social disorder, perhaps even a power vacuum. It is therefore understandable that some pre-Westphalian political philosophers decried mercenaries as inimical to good governance[10] and advocated state monopolies on coercive force as the antidote.

During the eighteenth century, European states began limiting their reliance on mercenaries in favor of creating their own standing militaries. Curtailing private forces was a vital step to enacting the Westphalian conception of sovereignty, which cast violent nonstate actors as criminal unless they were closely aligned with a particular patron. Technological development played a role in this process, with specialized handheld weapons that once favored mercenaries who had time to train in their use being supplanted by expensive artillery and mass-produced firearms that favored states capable of industrial production.[11] Ideology also played an important role in this process. The French Revolution marked a turning point "when national ideals were tied inextricably to military service."[12] A new sensibility of military participation as a civic responsibility emerged, with nations and their governments being forged through the unifying experience of war as an expression of solidarity and national self-determination. "As this concept of citizen duty to the state grew, and as patriotism and nationalism became increasingly seen as desirable and practical for armies, a selfish and financial motivation became mor-

10. Machiavelli is the most notable example. See Niccolo Machiavelli, *The Prince* (New York: Barnes and Noble, 2004); Niccolo Machiavelli, *The Art of War* (New York: Barnes and Noble, 2004).

11. Michael Mallett, *Mercenaries and Their Masters: Warfare in Renaissance Italy* (Barnsley: Pen & Sword, 2009); Charles Tilly, *Coercion, Capital, and European States, AD 990–1992* (Cambridge, MA: Wiley, 1993).

12. Kevin A. O'Brien, "PMCs, Myths and Mercenaries," *Global Policy Forum* (February 2000), 62, https://www.globalpolicy.org/component/content/article/173-sovereign/30555-pmcs-myths-and-mercenaries.html

ally inappropriate and practically inferior."[13] The nationalistic conception of war helped to reconstitute fighting as a unifying activity that brought people together and that could propel states to greatness. From this viewpoint, participation in war had to come from citizen soldiers. Mercenaries tarnished the civic obligation and risked corrupting war's progressive potential by reducing it to a commercial activity.

Before the twentieth century it was common for armed forces to rely heavily on private logistical support. Armies often marched with long trains of camp followers that provided for their needs, while ships purchased their supplies in the ports of allied and neutral countries.[14] Soldiers and sailors could focus on fighting, leaving the work of supply to civilians in most instances except when operating in inhospitable and unpopulated areas. During the nineteenth century, European armies gained speed by doing away with camp followers. Initially they allowed their soldiers to forage for provisions. This was one of Napoleon's contributions to logistics, and it was made possible by his strategic innovation of dispersing his forces across a large area along multiple lines of march (concentrated armies usually cannot find adequate provisions to live off the land for long).[15] Later in the century, the advent of railroads supplemented foraging with supplies sent from the home country. This granted greater resource security and allowed armies to grow in size. It was also a step toward the integration of logistics into the military profession.[16] During the wars of the nineteenth and twentieth centuries, armed forces expanded outward, taking over vital civilian occupations like factory production and rail transportation. This was in a sense the opposite of mercenarism, the intrusion of the military into private industry at the height of the Westphalian system.

The post–Cold War era marked a resurgent enthusiasm for privatized military force. On the supply side, this industry was made possible by the surplus of experienced personnel and weaponry available for the international market. Global demilitarization rendered more than seven million soldiers unemployed, with many lacking skills aside from those of war.[17]

---

13. Sarah Percy, *Mercenaries: The History of a Norm in International Relations* (New York: Oxford University Press, 2007), 12.

14. Van Creveld, *Supplying War*.

15. Gunther E. Rothenberg, *The Art of Warfare in the Age of Napoleon* (Indianapolis: Indiana University Press, 1978), 129.

16. Christian Wolmar, *Engines of War: How Wars Were Won and Lost on the Railways* (London: Atlantic Books, 2010).

17. "In global context, more than 7 million servicemen were thrust into the employment

Weak states and the nascent contractor organizations benefited immensely from the flood of cheap Soviet military hardware. Taken together, "this meant that both military expertise and equipment were in surplus supply at relatively modest prices on the open market."[18] On the demand side came the need to stabilize weak states that were either left with security crises as the two Cold War factions reduced support for foreign militaries or that were experiencing domestic conflict associated with decolonization. During the 1990s, armed contractors fought primarily in developing countries, but they established a foothold in more powerful armed forces, especially the US military, by taking over logistics work. Going into the twenty-first century, new demands emerged as strong states became embroiled in humanitarian and counterinsurgency missions that exceeded their capacities, especially when contending with post–Cold War size reductions.

The importance of contractors for powerful armed forces was vividly illustrated by the Iraq War. Private forces constituted the second largest part of the coalition, superseding the UK's commitment of 10,000 soldiers. By 2007, there were more than 100,000 US-employed contractors in Iraq, alongside around 160,000 uniformed combatants. Contractors consistently made up around 40 percent of the coalition.[19] Most of these personnel were involved in reconstruction projects, but many also specialized in combat activities that were beyond the scope of conventional forces. Thousands of civilian government officials and private companies were involved in reconstruction. They had the task of rebuilding Iraqi government and industry from the ground up, making it essential for them to work around the country. Given the general lack of security, civilians operating in Iraq turned to contractors for protection. American military forces were spread far too thin to take on the additional burden of guarding diplomats, and most were untrained for that kind of work.

Other armed forces likewise rely on private assistance. The UK is home to some of the world's largest private security firms, such as G4S, Aegis Defence Services, and Control Risks. It has hired contractors for operations in Iraq, Afghanistan, and Libya, as well as being a major contributor to counterpiracy

---

market." Lindy Heinecken, "Outsourcing Public Security: The Unforeseen Consequences for the Military Profession," *Armed Forces & Society* 40, no. 4 (2014): 625–46, 633.

18. Heinecken, "Outsourcing Public Security," 628.

19. James Jay Carafano, *Private Sector, Public Wars: Contractors in Combat—Afghanistan, Iraq, and Future Conflicts* (Santa Barbara, CA: Praeger, 2008); Bruce E. Stanley, *Outsourcing Security: Private Military Contractors and U.S. Foreign Policy* (Lincoln, NE: Potomac Books, 2015).

efforts.[20] G4S is one of the largest private employers in the world. With around 625,000 employees, it is substantially larger than the British military's roughly 200,000 personnel.[21] China has an estimated "5,000 private security companies . . . that employ almost 3 million security officers."[22] Security for its Belt and Road Initiative (BRI) is being largely provided by contractors, including Blackwater founder Erik Prince. Its contractors have spread globally in an attempt to protect Chinese workers and business interests abroad.[23]

Russian contractors are shrouded in secrecy. It is often unclear how closely they are linked to the government, and they are difficult to identify because of the Russian penchant for deploying masked soldiers who lack identifying insignia. However, there is evidence of Kremlin-backed contractors operating in the Central African Republic, Syria, Libya, and Ukraine.[24] These fighters have been instrumental in allowing Russia to project power while maintaining a degree of plausible deniability. For example, members of the secretive Wagner Group entered Crimea during the early days of the dispute in 2014 and lent support to local fighters to create the impression of a domestic uprising against the Ukrainian government. Private contractors were then sent into the Donbas region to repeat the process. By giving the impression that local actors are responsible for the violence against Ukrainian government forces, contractors have allowed Russia to downplay the fact that it is involved in one of the most aggressive territorial seizures since the Second World War.

Professionalism and Profit

State armed forces classify their members as military professionals who have the special rights and responsibilities associated with that status.

---

20. https://www.theguardian.com/business/2016/feb/03/britain-g4s-at-centre-of-global-mercenary-industry-says-charity

21. Luke McKenna and Robert Johnson, "A Look at the World's Most Powerful Mercenary Armies," *Business Insider*, February 26, 2012.

22. Alessandro Arduino, "China's Private Army: Protecting the New Silk Road," *The Diplomat*, March 20, 2018, https://thediplomat.com/2018/03/chinas-private-army-protecting-the-new-silk-road/

23. Sergey Sukhankin, "Chinese Private Security Contractors," Jamestown Foundation, May 15, 2020, https://jamestown.org/program/chinese-private-security-contractors-new-trends-and-future-prospects/

24. Sabra Ayres, "Russia's Shadowy World of Military Contractors: Independent Mercenaries, or Working for the Kremlin?" *Los Angeles Times*, February 18, 2018, http://www.latimes.com/world/europe/la-fg-russia-mercenaries-20180218-story.html

Moreover, they cultivate a sense that the military is an exceptional profession, distinct from all others in its moral authority and relationship with a state. Martin Cook says that "Military personnel live in a unique moral world. They exist to serve the state. The essence and moral core of their service is to defend that state through the management and application of violence in defense of the territorial integrity, political sovereignty, and vital national interests of that state."[25] This sense of professionalism depends on militaries jealously guarding their status, maintaining high levels of proficiency, and imposing ethical constraints on members.[26] Carving out a special status for soldiers helps to imbue their work with a redemptive meaning that justifies killing and self-sacrifice.

Contractors aggravate the problem of defining professional membership. On the most basic level, they challenge the uniqueness of military expertise, as "the military is no longer the sole provider of military knowledge, but has become a consumer thereof."[27] The multiplicity of private forces suggests that martial skills and the authorization to use them are shared qualities. By extension, contractors threaten ostensibly unique values that are said to characterize that profession, values designed to guarantee good conduct. George Lucas rightly argues that "professional ethics in a military context remains vexed, contested, and not well understood by the public."[28] Military ethics doctrine and training often revolves around efforts to define the profession, under the assumption that doing this can identify the ethical identity unique to military professionals. Lucas goes on to argue that "If we take the problem of uncertainty and disagreement over professional ethics and military virtues out of its public context and place it instead in the private sector, we generate even more difficulties."[29] That is, apart from anything the contractors do, merely posing an alternative form of military power undermines efforts to foster an identity that is unique to state military forces and to build an ethical awareness around that identity.

Contractors are not bound by the same accountability mechanisms and ethical codes as ordinary soldiers, nor are they as clearly subject to domestic or international laws. In dozens of instances, contractors have been able to escape punishment or even investigation for serious crimes, in some instances

25. Martin L. Cook, *The Moral Warrior: Ethics and Service in the U.S. Military* (Albany: State University of New York Press, 2004), 123.

26. Nicholas Fotion and Gerard Elfstrom, *Military Ethics: Guidelines for Peace and War* (New York: Routledge, 1986), 47.

27. Heinecken, "Outsourcing Public Security," 7.

28. Lucas, *Military Ethics: What Everyone Needs to Know*, 136.

29. Lucas, *Military Ethics What Everyone Needs to Know*, 136.

crimes that soldiers were ultimately punished for. "In 2002, for example, civilian contractors working for DynCorp, a security and equipment maintenance firm, were accused of keeping a ring of young women and girls as sex slaves in Bosnia. There was no investigation; instead, the alleged perpetrators were simply fired and sent home."[30] The soldiers involved in the Abu Ghraib prisoner abuses were brought to trial, court martialed, and imprisoned.[31] By contrast the contractors implicated were able to escape punishment. Incidents of misconduct add to the sense that contractors are unethical combatants who tarnish the reputation of state military forces.

Those who defend contractors argue that the threat to military professionalism and limited accountability are transitory problems that exist because of the novelty of the contractor market, and that stricter regulations are apt to take shape in the coming years.[32] It is reasonable to believe that existing conditions may change in the future and that contractors could in principle obey the same rules as regular armed forces. However, this is ultimately an unsatisfying response because it leaves too much to chance. International markets are notoriously difficult to regulate, especially when companies are able to shift operations from one place to another. Tax evasion, the movement of operations to areas with weaker environmental and labor regulations, and efforts to cultivate unethical influence with government officials are rampant. States strive to attract large investments from the world's most powerful companies by offering incentives, such as reduced (or nonexistent) taxation and exemption from regulations. It is plausible that contractors who are vital to national security could receive similar treatment. I agree with former contractor Sean McFate's assessment that "if trends of the past are indicators of the future, then the new market will be lightly regulated, and efforts by countries to change that will simply drive PMCs offshore or underground."[33]

Claiming that contractors can be brought under the same rules as conventional military forces seriously underestimates the practical difficulties associated with this task. Existing research has yet to establish a convincing model of how regulations that are enforceable for monopolized state mili-

30. Anni Baker, *Life in the U.S. Armed Forces: (Not) Just Another Job* (London: Praeger, 2008), 86.

31. Baker, *Life in the U.S. Armed Forces*, 87.

32. Uwe Steinhoff, "Ethics and Mercenaries." In *New Wars and New Soldiers: Military Ethics in the Contemporary World*, ed. Jessica Wolfendale and Paolo Tripodi (Burlington, VT: Ashgate, 2011), 137–52.

33. Sean McFate, *The Modern Mercenary: Private Armies and What They Mean for World Order* (Oxford: Oxford University Press, 2017), 4.

tary forces could ever be applied to international firms, except to provide the classic free market response that states should discourage bad practices by simply not forming contracts with unethical entities. Such ethical self-restraint is impossible if a state has become reliant on external military assistance and has placed itself in a position of being unable to mobilize sufficient domestic military reserves during a crisis. Armed forces are expensive. The costs of military preparedness cut into domestic spending and may be electorally unpopular. Countries that lack the resources to maintain large standing militaries without a clear security rationale may be tempted to forgo this expense during peacetime. The availability of contractors to assist in interventions raises the prospect of being able to hire military assistance if and when it is needed. This makes it tempting to avoid maintenance costs while still being able to act in self-defense when compelled to do so. If a state places itself in this position, it will have little choice about employing contractors during an emergency, regardless of whether the firms in question have a record of good conduct.

Singer shows that weak states can easily fall into a trap by pursuing this strategy. States may forgo investing in their own security, then rush to hire contractors each time a threat arises, only to find that "client dependency grows each time they outsource or privatize functions. The client loses expertise and capabilities and becomes more reliant on the PMF [private military firm]." In the end, client states may become trapped in a vicious cycle of dependency that lays the foundations for long-term security instabilities.[34] From an international perspective, cycles of dependency preclude the prohibition of contractors, as any serious restriction on them from international regulatory agencies could leave client states without any substantive protection. Assuming that these commentators are right in thinking that contractors offer greater effectiveness, employing them increases the dependency of the hiring state and decreases their leverage over the contractors.[35] This magnifies the challenge to sovereign authority. It likewise erodes norms by making it more challenging for states to monitor and enforce compliance with them.

On the contractor side, there are powerful incentives to avoid accepting any new regulations that could fill the normative vacuum. Corporations have a duty to their shareholders to externalize costs whenever possible and to

34. Singer, *Corporate Warriors*, 78.

35. Doug Brooks and Matan Chorev, "Ruthless Humanitarianism: Why Marginalizing Private Peacekeeping Kills People," in *Private Military and Security Companies: Ethics, Policies and Civil-Military Relations*, ed. A. Alexandra, D-P. Baker, and M. Caparini (New York: Routledge, 2008), 116–30.

circumvent regulations that would damage their financial interests. "The newest wave of private military agents are commercial enterprises first and foremost."[36] They are organized like businesses, profit driven, compete in an open market, and have links to corporations and financial markets. New regulations on contractors will not change this fundamental incentive problem, and the regulations may well be impossible when contractors can remain in business by shifting operations to less restrictive patrons. Moreover, they have ample freedom to do this because conflicts usually take place in areas without effective legal institutions and where competitors are apt to overlook norms because their survival is at stake.

The profit motive is also problematic in its own right, beyond any impact this might have on regulatory oversight. The use of contractors by states was a logical upshot of the turn away from conscription to all-volunteer forces in many of the world's leading militaries throughout the 1960s and 1970s. Military service went from being a shared civic obligation to a dangerous, low-paying occupation that most people wanted to avoid. It retained a certain allure and prestige and continues to be widely regarded as a mark of profound patriotism and devotion. Nevertheless, the rise of all-volunteer militaries helped to break the sense of military service as a kind of universal civic obligation, and instead made it a career choice that potential recruits would weigh against competing opportunities. This shifted membership in the armed forces from the public sphere into the private sphere, pushing armed forces into the labor market to attract recruits on similar terms as other employers. Contractors can enter this market to offer better pay and benefits, while also cutting their own training and acquisition costs by hiring veterans who have received extensive regular military training. The pursuit of profit may be the single greatest concern for those who object to military privatization. Coady captures this sentiment perfectly when he argues that "someone who hires his gun to the highest bidder or, less dramatically, fights predominantly for money will typically lack the motive appropriate to war."[37]

The fact that contractors often make more money than regular soldiers is cause for considerable resentment, which adds to the threat contractors pose to the identity of professional soldiers.[38] One former US Marine encapsulates this sense of difference when he says that "They weren't one of us, one of

---

36. Singer, *Corporate Warriors*, 45.

37. C. A. J. Coady, "Mercenary Morality," in *International Law and Armed Conflict*, ed. A. G. D. Bradney (Stuttgart: Franz Steiner Verlag, 1992), 55–69, 63.

38. Marcus Schulzke, *Pursuing Moral Warfare: Ethical Theory and Practice in Counterinsurgency Operations* (Washington, DC: Georgetown University Press, 2019); Scahill, *Blackwater*.

the troops. They were mercenaries—soldiers of fortune. They were doing it for the dough. . . . It was unsettling to hear a man talk about money as he packed off to war. It kind of took the nobility out of it."[39] A British officer describes them as "a dark and sinister conglomerate whose tentacles wrapped around every tender spot the army presented to it, and suck its service supply budgets while offering the meagrest possible service in return."[40] Whether or not we can find any deeper immorality in contractors, this antipathy from military personnel should encourage us to confront the risks associated with aggravating the identity crisis affecting the military profession. It shows that the existing norms of professional militaries are intact to an extent, but that they are clashing with the realities of increased privatization and causing division between ostensibly allied personnel.

Steinhoff defends contractors by saying that "there is no good reason to suppose that mercenaries are more likely than regular soldiers to fight in unjust wars. There is no good reason, either, to suppose that their fighting in a war is driven by stronger monetary motives or intentions than in the case of regular soldiers."[41] In fact, he argues that it is actually "much easier for a mercenary to avoid participation in an unjust war than for a member of a regular force."[42] Because contractors have much greater control over when they will fight, the reasoning goes, they have the freedom to make fairly disinterested decisions and can forgo fighting if a war is morally objectionable. It is likely that Steinhoff overstates this freedom. After all, contractors must make money somehow, and they cannot make money if they refuse to fight. They must participate in their industry just as any other business would, and they put themselves in jeopardy by refusing jobs based on moral grounds. Forgoing one war would make a firm less attractive to other potential employers, especially belligerents who would not want the public relations embarrassment of having their morals questioned by contractors. Given the pervasive disagreements over when wars are justified, it is doubtful that a major firm could remain competitive if it stayed out of every morally dubious conflict. Thus, the threat contractors pose to existing norms of war and the military profession is not likely to be counterbalanced by any new morality of professional ethics emerging from the free market.

39. Tyler E. Boudreau, *Packing Inferno: The Unmaking of a Marine* (Port Townsend, WA: Feral House, 2008), 63.

40. Graham Lee, *Fighting Season* (London: Duckworth Overlook, 2012), 107.

41. Steinhoff, "Ethics and Mercenaries," 140.

42. Steinhoff, "Ethics and Mercenaries," 140.

Recriminations and Rebranding

Fighting for profit may arguably not be unethical, first, because many workers do unpleasant jobs for money and, second, because contractors can have noble motivations aside from profit.[43] However, this does not change the fact that the mere perception that wars involving contractors are being waged for profit strikes a severe blow to the status of war as a social practice imbued with a sense of nobility. Concern over the corrupting influence of profit motives extends beyond the challenge of defining the military profession to become a more general political problem having to do with how armed forces operate as agents of their societies. Cathy Downes argues that the military depends on the image that members are inspired by a sense of selfless service. Without this, "society would be loath to allow the military profession to retain its monopoly over the profession."[44] This was borne out during the Iraq War, when contractors were widely decried by journalists and activists. Critics raised many arguments against contractors, but foremost among them was a sense of wrongness about commercializing war.[45] And this in turn aggravated concerns that the war was being waged because of private greed on the part of business interests within the government.

The unabashed instrumentalism of fighting for money, rather than for the nationalistic, ethnic, or religious motives that constitute war as a collective obligation, is deeply unsettling for people who have grown up in a culture that has attempted to give war meaning by appealing to these beneficent motives. Colin McInnes argues that "by making war the business of professionals, Western societies absolve themselves from some of the responsibility of placing soldiers in harm's way."[46] The privatization of war strips away an important part of that existential meaning that redeems violence, reducing conflict to a more purely instrumental affair akin to market transactions. When war takes place for profit and without a sense of these overriding value systems, it is reduced to pure political and commercial machinations. The killing becomes far more difficult to incorporate into a legitimizing ideology. It is ironic that contractors rose to international prominence during the war in Iraq, as the rebuilding effort depended on unifying the country with a new nationalistic ideology that could transcend ethnic divisions. That is to say,

43. Steinhoff, "Ethics and Mercenaries," 140.

44. Cathy Downes, "To Be Or Not to Be a Profession: The Military Case," *Defence Analysis* 1, no. 3 (1985): 147–71, 159.

45. Scahill, *Blackwater*.

46. Colin McInnes, *Spectator Sport War: The West and Contemporary Conflict* (Boulder: Lynne Rienner, 2002), 69.

the coalition became involved in imparting the same kind of nationalistic ideology that it was actively undermining by employing contractors. At times contractors were even tasked with training local forces and imbuing them with this sense of patriotic devotion that was implicitly contradicted by their commercialization of war.[47]

Furthermore, contractors may inflict a high perceptual cost when it comes to the civilian audience within the contested area. Contractors may not be part of the military, but they are widely perceived as such by those civilians. Any infractions by individual contractor firms may be generalized to the security forces more broadly. Armed forces are already at pains managing their images in the struggle to "win hearts and minds"; their efforts are further complicated when they must be answerable for how fighters outside their chain of command behave. The formal structure of military professionalism therefore has profound implications for the perception of war, and especially for risk-management strategies that are rooted in perception management. Militaries rely on private firms to support the broad spectrum of operations essential to counterinsurgency even as certain contractors undermine the strategic and political goals with high-profile acts of misconduct.

Contractor firms engage in their own perception management to distance themselves from incidents and to contest the deeply entrenched norm against fighting for pay. They must also market their services to potential employers and attract the most skilled military veterans as new employees. "Since the barriers to entry are so low, corporate branding and reputation are the keys to moving into a position of dominance in the market."[48] None have demonstrated this more forcefully than Blackwater, which undertook a branding campaign that extended into clothing, military equipment, and even video games. Founder Erik Prince describes seeing other military contractors and civilians buying the Blackwater clothing and wearing it in contested areas. Prince recalls that during the Iraq War, "Suddenly our bear paw logo was everywhere—on T-shirts and baseball caps worn by everyday security guards, Iraqi civilians, and ironically everyone but my employees, who hardly ever dressed in branded gear."[49] They were able to license their image and promote their brand even as their own personnel avoided marking their membership in the group because of their effort to maintain a low profile.

47. One of the best examples of this comes from DynCorp, which was tasked with training Iraq's national police force.

48. Singer, *Corporate Warriors*, 74.

49. Erik Prince, *Civilian Warriors: The Inside Story of Blackwater and the Unsung Heroes of the War on Terror* (New York: Penguin, 2013), 82.

The public relations campaign ultimately failed to save Blackwater's image, but the organization survived in typical corporate fashion by rebranding itself first as XE Services, then Academi.

Few contractors can match Blackwater's public relations campaigns, but most major firms strive to cultivate a positive appearance. Their websites tend to focus on humanitarian missions rather than combat. They use corporate language about offering "consultancy," "insight," and "efficiency." Rebranding and reorganization are also common. Executive Outcomes disbanded in 1998 in response to South Africa's Regulation of Foreign Military Assistance Act, which would increase government oversight by requiring authorization of all contracts with foreign governments.[50] However, the remnants formed new companies that continue to operate. Blackwater has changed names multiple times, as have Sandline and Triple Canopy. The names the companies choose are themselves revealing. These usually incorporate some vaguely corporate language, such as Academi, DynCorp, or Aegis Defence, that shift focus away from the use of force. Taken as a whole, these perception-management techniques show that contractors attempt to redefine their activities and themselves as existing beyond the boundaries of war.

State armed forces are likewise guilty of manipulating language, so this style of perception management is by no means unique to contractors, but the context for them is much different because contractors are competing in a market and striving to shape perceptions to advance their companies' financial interests. State militaries attempt to legitimize themselves by appealing to patriotic motives, and above all loyalty to the state. This reinforcement of existing political institutions gives them a conservative character. Contractors, by contrast, face an uphill battle to legitimize themselves and cannot do this on the same terms as regular militaries. They cannot appeal to patriotism and selfless service to the state and so must find alternative justifications for violence. Some goals, like saving innocent people and protecting international commerce, may be noble. However, regardless of whether these legitimizing missions are worthwhile, they introduce new reasons for fighting apart from those regular forces appeal to, thereby changing the rationale for political violence and further confusing attempts to delineate clear boundaries of war and peace.

50. https://www.globalsecurity.org/military/world/para/executive-outcomes.htm

Contractors and Risk

For hiring states, perception management is a kind of risk management insofar as reliance on contractors can reduce oversight and diminish the likelihood of public backlash against the use of force. The United States' dependence on contractors throughout the War on Terror helped to sustain a narrative of minimal US military commitments. Starting with the invasion of Afghanistan, which initially only involved around three hundred fighters, and continuing with an invasion of Iraq that involved around one-fifth the number of the soldiers who went into Kuwait a decade earlier, the Bush administration pursued a variant of the revolution in military affairs sometimes called the "Rumsfeld Doctrine." Lowering the price, both human and financial, of military commitments was necessary for demonstrating that the country could meet its aspiration of fighting terrorism around the world. Contractors formed an essential component of this approach, as they provided the combat and logistical strengths to offset weak support from partners in the coalition. They greatly magnified military power while still feeding "a false impression of a much smaller U.S. military footprint and national commitment."[51] News reports on those conflicts persistently omitted contractor numbers from estimates of the overall deployments as well as from casualty figures, which contributed to a broader narrative of those being limited wars waged cheaply and with little loss of life.

Contractors have aided Russian perception management efforts in at least two ways. First, they have been instrumental in attempts to hide the extent of assistance being given to foreign fighters. For example, the Wagner Group helped the Syrian army recapture Palmyra in 2016 without giving the impression of Russian involvement.[52] In 2018, a sustained battle between the Wagner Group and a Kurdish militia force at Deir Ezzor featured American soldiers and Russian contractors on opposing sides. It is likely that this would have provoked a major international incident if not for the Russian government's ability to claim that the contractors were not acting under orders. Second, using contractors to provide assistance to local fighters, as Russia did in Ukraine, is a way of concealing the aggressive seizure of territory. In this

51. Micah Zenko, "The New Unknown Soldiers of Afghanistan and Iraq," *Foreign Policy*, May 29, 2015.

52. Sergey Sukhankin, "Russian PMCs in the Syrian Civil War," Jamestown Foundation, December 18, 2019, https://jamestown.org/program/russian-pmcs-in-the-syrian-civil-war-from-slavonic-corps-to-wagner-group-and-beyond/

sense, contractors are a loophole in the Westphalian norm that such aggression constitutes an act of war between states.

Equally important is that contractors help to maintain the impression of relatively risk-free military operations. Throughout the preceding chapters I have explored a few of the many ways in which armed forces have attempted to limit their casualties, especially deaths. War is being remade as a less costly activity that can continue in an era when many policymakers fear that public support will erode as the death toll rises. By relying on help from contractors, states can downplay the overall casualty numbers. Contractor losses do not have to be included in official estimates. The extent of their losses may not even be known by the governments that employ them. The United States is unable to determine exactly how many contractors have been killed in its service.[53] The fragmentation of command and control facilitates this by making it difficult to establish any effective oversight even for civilian government officials and members of the military. Contractors are likewise not required to disclose information through mechanisms such as the US or UK Freedom of Information Act. Even if oversight improves, it is unlikely that this accountability gap would ever be fully overcome because contractors are nonstate entities with their own accounting procedures that may be checked, but not controlled by, their employers. In those instances when individual companies are forced to disclose information or submit to stricter oversight, they are quick to reorganize under a new label or hire subcontractors who are not themselves bound by the rules.

Deploying private forces to a conflict does not entail the same level of commitment as deploying regular military forces, nor is it likely to be perceived that way. Contractors can be hired in small numbers, for fixed terms of service, without requiring the sustained investment that follows conventional military operations. This flexibility is useful in an era when conflicts are not resolved but only managed with the goal of making them less threatening in the short term. "If problems have to be managed rather than definitely solved, then the most important political objective is to maintain freedom of action. Knowing that new risks will arise, a government needs to retain resources to deal with the next problems, and the next, and the next."[54] This offers further explanation of why contractors are attractive for the

53. Office of the Special Inspector General for Iraq Reconstruction, *SIGIR Special Report Number 2: The Human Toll of Reconstruction or Stabilization during Iraqi Freedom* (2012).

54. Mikkel Vedby Rasmussen, *The Risk Society at War: Terror, Technology and Strategy in the Twenty-First Century* (New York: Cambridge University Press, 2006), 36.

American and British militaries. Those countries have considerable military strength but are continually waging small wars around the world and are at risk of being overextended unless some of the burdens can be shifted onto others. The War on Terror in particular is the kind of intractable, global struggle that cannot be prosecuted by any single belligerent. Threats rise and fall; they move from country to country. It is easier to pursue such threats when state military forces are able to emulate enemies by launching small operations worldwide.

One danger in terms of how contractors shape perceptions of armed conflict is that they may help to eliminate the perception that wars are being waged at all. Münkler argues that "Mercenaries unlike professional soldiers or conscripts originating in the national electorate, have no possibility of eliciting a political response in the event of high-risk and high-loss operations."[55] This claim that there is no possibility of public backlash is certainly too strong, as contractors have received considerable opposition from journalists, academics, and concerned citizens. The press surrounding contractors is frequently negative.[56] Nevertheless, the underlying concern is valid. Contractors may not always escape public scrutiny, but they certainly increase the prospects of doing so successfully. Contractors are routinely involved in small-scale confrontations in areas that receive little media coverage, such as those in Nigeria, Mali, and the Central African Republic.

By outsourcing certain military roles, politicians can improve their capacities for managing the perceived risks associated with war and distance themselves from the resulting casualties. The shifting of responsibility is a dynamic process that changes over time and is countered by attempts to increase oversight and journalists' reports on crimes committed by the contractors. New oversight mechanisms could potentially be created, but the challenges of doing this are greater than for conventional militaries. Moreover, with the world's leading armed forces benefiting from oversight loopholes, it is doubtful that they will be eager to improve public accountability. Contractors and their employers respond to heightened scrutiny with new strategies for managing perceptions and evading responsibility. The most notable trend here has been toward greater reliance on foreign fighters. As of 2016, US government reports indicate that only around 40 percent of contractors in the US Central Command Area of Operations were American citizens and that they were suffering higher casualties than American soldiers.[57] This points to a

55. Herfried Münkler, *The New Wars* (Malden, MA: Polity Press, 2005).

56. For example, see Scahill, *Blackwater*.

57. https://www.acq.osd.mil/log/ps/.CENTCOM_reports.html/5A_July_2016_Final.pdf

future in which conflict is not only privatized but also carried out by people who have few personal ties to the belligerents they serve. McFate, who has personal experience in the industry, says that "It is certainly conceivable that a PMC today could secretly and simultaneously serve two clients at war with each other."[58] He even predicts that contractors with no political stake in conflicts could lengthen wars to maximize earnings. In this case the risk-management gains for relying on contractors could be undercut by the new risks associated with prosecuting endless wars with mounting costs.

Fragmentation of the Battlespace

Contractors acting on behalf of states are not in the military chain of command, but they must be integrated into military operations and the overarching political strategy. This is challenging, as structural differences between private organizations and the military foster miscommunications and disagreement. Soldiers and contractors may be on the same side, but they act through distinct channels and come into conflict about mission priorities and procedures.[59] Moreover, contractor firms are extremely diverse. Each has its own internal structure, culture, and objectives. There is no single model that all follow. They likewise have compelling fiduciary and financial interests distinct from those of allies operating in the same area. For example, contractors hired to guard an American diplomat in Baghdad may be unwilling to assist nearby American soldiers who are under attack. When dozens of contractor firms operate in the same area, coordination problems are routine. The result is fragmentation of the battlespace. Rather than a battlespace made up of friends and enemies—or even a battlespace made up of friends, enemies, and civilians—there is a multipolar battlespace made up of heterogeneous coalitions. The many different actors on either side participate in war as distinct entities with semi-independent goals. This takes the fighting beyond the relatively clear structure of conventional war.

Of course, contractors do not always share the battlespace with soldiers. They are not geographically confined in the same ways as regular armed forces. States are, at least in principle, only supposed to fight in self-defense or when defending a victim of aggression. This restricts the scope of violence to areas where some kind of territorial defense or appropriation is taking place. Furthermore, armed forces are usually only deployed when there are compelling national interests for doing so. States sometimes intervene to pre-

58. McFate, *The Modern Mercenary*, 1.

59. For examples, see Scahill, *Blackwater*.

vent humanitarian crises, but they do so inconsistently and usually with clear geographical preferences. Failure to stop the killings in Rwanda, Sierra Leone, and Darfur when similar interventions were launched in the Balkans and oil-rich regions provides ample evidence of this. Even as counterinsurgency operations have done away with front lines and made conflicts more fluid, militaries remain preoccupied with securing areas, conducting patrols, and projecting power spatially.

Because they go wherever employers are prepared to pay for their services and lack the need to justify their missions in terms of the logics of territorial self-defense or national interest, contractors are much less geographically constrained than are conventional forces. They rarely operate according to the same Westphalian logic of capturing and holding ground. Contractors are more often engaged in mobile operations, like guarding convoys and protecting diplomats. They do not fight for land but to protect infrastructure and people. The most prominent firm of the 1990s, Executive Outcomes, built its reputation by participating in African wars that European and North American states would not get involved in. They carved out a niche for themselves by taking their services to geographically unpopular areas for military interventions and migrating from war to war. In the twenty-first century, contractors have transformed into transnational companies carrying out operations around the world. For example, Constellis, heir to Blackwater and Triple Canopy, claims to operate in more than forty countries.[60]

Many of the organizations hiring private security are nonstate actors that lack geographical attachments of their own, at least to the extent that a state would have. For example, shipping companies traveling through areas with high rates of piracy hire guards to deter attack, and firms engaged in resource extraction purchase private assistance to protect employees from kidnapping. This involvement with companies that are themselves highly mobile compounds contractors' transcendence of the territorial component of conventional war. Moreover, it shows the extent to which states have lost their Westphalian monopoly on coercive force. It is not merely that contractors are sharing the security role with state forces but also that companies with no security function can enter the security field by purchasing contractors' services.

Free markets make it possible for goods and services to flow far more than when supply is regulated. Even if free markets are more efficient at allocating resources, they are mercurial; they follow boom and bust cycles as resources, jobs, and investments shift from one area to another. The global marketplace

60. https://constellis.com/who-we-are/global-footprint

has been the chief driver of globalization and is one of the central causes of instability by linking countries together in a shared economic system that can be destabilized by disruption to any of its components. Privatization of military force paves the way for a globalization of military force operating alongside an already huge international marketplace for weapons. In this context, destabilization has the potential to expand beyond economic activities narrowly construed to include issues that are vital to national security. As Singer points out, "The new privatized military industry also means that state and non-state actors alike now can access military capabilities formerly exclusive to strong states."[61] This will not completely disrupt geopolitical realities. Wealthy states that currently field the world's most advanced militaries have greater financial resources with which to hire private forces. Nevertheless, a whole range of moderately wealthy powers with small armed forces gain the ability to selectively participate in military affairs without having to build domestic institutions for that purpose. This change in the capacity to employ coercive force breaks coercive power free from its existing geographical constraints. More importantly, it makes geographical constraints as such less important. When it comes to private contractors, the real source of power is not in any particular location but in the companies themselves. Just as in other industries, they have the freedom to become transnational entities that can migrate at will.

Outside the Law

Private military contractors constitute a serious challenge to the longstanding Westphalian norm of states being the sole possessors of legitimate military power, a norm so deeply entrenched that this is often treated as the defining characteristic of the Westphalian state.[62] Contractors mark a return to the premodern understanding of military force as being detached from a national community and wielded for private purposes. The reliance on contractors in recent conflicts sets a precedent, which will be difficult to break away from as the complacency of following along a set track spreads. Krahman argues that we have already passed the point where this trend might be easily resisted, saying that "The failure of the international community to agree on a global regulation of PMSCs puts into question

61. Singer, *Corporate Warriors*, 70.

62. Max Weber, "Politics as a Vocation," in *The Vocational Lectures*, ed. David Owen and Tracy Strong (Indianapolis: Hackett, 2004).

whether the evolution of the norm toward an inclusion of state-sanctioned or regulated private armed force has already reached the phase of international 'norm internalization.'"[63] A new permissiveness toward private forces is taking shape, and it is doing so without adequate regulatory frameworks. Losing such a powerful normative constraint on war is regrettable, as informal norms can play an important part in limiting the scope and intensity of hostilities.

Contractors also raise legal problems because they do not fit neatly within international humanitarian law. That body of law emerged during the twentieth century and primarily addressed challenges that existed during this time. Although it borrowed from just war theory, it generally did so to reflect on specific problems arising during twentieth-century conflicts, such as which new weapons should be prohibited, how prisoners should be treated, and how wars could be justified in terms of self-defense. This framing was an understandable necessity of formulating clear laws that could be used to prevent the recurrence of atrocities, but it resulted in a body of law that was poorly suited for addressing future problems and that lacked enough explanation of general principles for them to be applied against all of the problems that would emerge later.

One of the foremost moral and legal challenges is that contractors are difficult to categorize in terms of the combatant vs. civilian binary, either as this is defined in the law or in prominent accounts of just war theory. Contractors themselves often resist the "combatant" label and obscure their status. Erik Prince, the founder of Blackwater, is one of the world's most prominent contractors and encapsulates this idea in the title of his book *Civilian Warriors*. This name captures the attitude that it is possible to be both a civilian and a warrior without contradiction. The label gives contractors some of the innocence of civilians by making them appear as reluctant fighters, while the warrior identity captures the contractors' efforts to present an ethos of military expertise. Claiming both identities at once undermines the distinction between the two and, along with it, challenges the binary that is at the heart of international humanitarian law. Without this distinction the project of regulating war becomes infinitely more difficult, both because it erodes civilians' protected status and because it raises questions about what rights and responsibilities should be associated with warriors.

The threat to civilian status is even more serious when it comes to how contractors conduct themselves in practice. Despite heavily criticizing Iraqi

63. Elke Krahman, "The United States, PMSCs and the State Monopoly on Violence: Leading the Way Towards Norm Change," *Security Dialogue* 44, no. 1 (2013): 53–71, 66.

insurgents for their attempts to blend into civilian populations, Prince admits that he and his Blackwater employees routinely did this. For example, he says that "I usually tried to blend in by traveling in standard dinged-up civilian cars."[64] Steve Fainaru reports seeing contractors in Iraq driving civilian vehicles with few visible signs of military function.[65] Robert Young Petton observes of DynCorp contractors in Afghanistan that "some dress OGA-style with distinctive safari vests and outdoor runners, while others sport a casual look that almost says fishing or weekend hunting."[66] Then there is the language surrounding these people, such as Prince's claim that they are "civilian warriors." Even journalistic reports of operations frequently describe these combatants as "civilian contractors" to differentiate them from uniformed soldiers.[67]

The use of contractors in support roles marks an important deviation from practices that upheld the principle of discrimination. Among just war theorists and members of the military, it is generally accepted that field support personnel located behind the lines are legitimate targets.[68] This is a reasonable view that was formed at a time of conventional warfare when an array of logistical operations, such as maintenance, supply, sanitation, cooking, and transportation were handled by uniformed soldiers who had some degree of combat training. Although they might be unarmed at the moment of attack, soldiers in support roles were nevertheless participants in war who were integral to the overall functioning of the military. They were likewise potential combatants who took up arms under extraordinary circumstances. This happened repeatedly during the Second World War when units were under sustained attacks and forced to defend themselves by any means available.[69] Similar situations can even arise in smaller operations, as when American support personnel took part in the Battle of Mogadishu in 1993.[70]

By outsourcing support roles to private contractors, states transform military roles into civilian roles or at least quasi-civilian roles. They can now

64. Prince, *Civilian Warriors*, 136 (on insurgents blending in); quotation on 156.

65. Steve Fainaru, *Big Boy Rules: America's Mercenaries Fighting in Iraq* (Philadelphia: Da Capo Press, 2008), 76.

66. Robert Young Petton, *Licensed to Kill: Privatizing the War on Terror* (New York: Three Rivers Press, 2006), 79.

67. Scahill, *Blackwater*, 210.

68. Marcus Schulzke, "The Morality of Remote Warfare: Against the Asymmetry Objection to Remote Weaponry," *Political Studies* 64, no. 1 (2016): 90–105.

69. For example, see Antony Beevor, *Ardennes 1944: The Battle of the Bulge* (New York: Penguin, 2016), 159, 232, and 281.

70. Mark Bowden, *Black Hawk Down: A Story of Modern War* (New York: Atlantic Monthly Press, 1999).

decry attacks on civilian support personnel who perform roles that once belonged to combatants as a violation of the principle of civilian immunity, thereby making it easier to frame adversaries' actions as illegitimate. Worse still, this civilianization of military support comes at a time when the desperation to break up terrorist and insurgent networks has led states to focus their attention on enemy leaders and support personnel. The result is a double standard when it comes to targeting practices, whereby a large segment of the participants in states' war efforts are unilaterally declared off limits while states expand their own targeting. Restriction on the scope of war would ordinarily be a welcome portent of reduced violence, but in this case it threatens to compromise the validity of the normative constraints on war because it so clearly favors states that are able to employ contractors.

It is important to recognize that these concurrent normative problems go beyond contractors themselves. By employing private forces, states change the character of their own actions. They cannot help but share in the consequences of a style of warfare they have helped to produce. "Through privatization, the state's agent of action is now no longer its national military, but instead a profit-motivated actor."[71] States render themselves dependent on contractors and abdicate their power as institutions monopolizing the use of legitimate force. That is to say, states are willfully endangering their longer-term status in international politics by relying so heavily on nonstate actors, and they are doing so for the short-term gains of waging wars more cheaply and with less visible losses. They are likewise endangering the status of critical norms restricting war and limiting violence against civilians.

The Private Individual

Contractors are inherently individualistic, especially when seen in contrast with regular military forces, because they are free market actors. Like corporations, military firms pursue profit. Effectiveness matters too, yet only because of the need to secure future work. The companies rarely have a direct stake in any particular political outcome, except insofar as they must support their employer. Their top long-term priority is to ensure their own profitability. They may take on noble goals in some instances, such as by helping to end genocides or fighting alongside state militaries defending their homeland, but they do this instrumentally, as a calculated decision to ensure success in the market. As Singer points out, "With PMFs, clear ten-

71. Singer, *Corporate Warriors*, 170.

sions always exist between the security goals of clients and the firms' desire for profit maximization."[72] They even perversely have an incentive to avoid ending wars. McFate finds that contractors are apt to stretch fighting out as long as possible, even escalating it, to ensure that they retain employment.[73] Given these incentives, the firms involved in a given conflict may pursue divergent goals that correspond to narrow agendas.

The members of contractor firms are likewise often treated as individuals first, rather than as the collective members of a profession at arms or as the collective of patriotic defenders of homeland embodied by the regular military. Firms appeal to individuals' profit-seeking behavior and entice them with financial rewards instead of moral rewards. Members are not supposed to join because of any loyalty to the organization itself. There is no patriotic attachment to Blackwater, DynCorp, or Academi. Rather than appealing to collective ideals, the firms recruit based on appeals to the self-interest of members. Above all, there is the promise of lucrative jobs. Thus there is an individualistic incentive structure both at the level of the organization and for each person in the organization, which in both cases is based on financial rewards. This trend toward greater individuation at the macro and micro levels means increased discretion over when agents (as either groups or individuals) participate in wars.

Individuation could arguably be advantageous. Defenders of private military forces tend to think that combatants should have more freedom to select how and when they fight and that this discretion would help to limit the incidence of unjust wars.[74] However, this is a dubious proposition that makes the mistake of assuming that contractors as organizations or individuals would be primarily motivated by the nature of the conflict rather than the profit they can make by fighting. The Iraq War was widely condemned as being unjust and aggressive, and yet there was little trouble recruiting tens of thousands of private contractors to participate. The capture of Crimea and the Donbas went against the principles of *jus ad bellum*, and yet the Russian government was able to find willing participants. We should face the facts that the morality of a conflict may be *among* the considerations that contractors use to decide when they will fight, but that contractors are by no means charitable actors and that this one motive is not likely to take precedence over others. There is no more reason to think that contractors will pursue just wars than there is to assume that other types of corporations will unilaterally

72. Singer, *Corporate Warriors*, 151.
73. McFate, *The Modern Mercenary*.
74. Steinhoff, "Ethics and Mercenaries," 137–52.

attempt to promote equitable practices within their industries. Even when companies do this, it usually seems to be a calculated move to win support and a greater market share rather than being a purely benevolent decision. When lives are on the line, the hope that contractors will take on only the most justifiable military operations is unrealistic. Virtually everything we know about the behavior of free market actors suggests that they are highly individualistic and apt to externalize costs whenever possible, even at the expense of high moral, environmental, and economic costs.

To some extent, the contractor market exists because of broader trends toward individuation in twenty-first-century warfare that run throughout this book. One of the defining contractor jobs is guarding civilian government officials in contested areas. Counterinsurgencies have increased the demand for diplomats, aid workers, and engineers to work in contested areas, as they are essential for rebuilding infrastructure and shattered political institutions. Local politicians likewise have high value and are at risk of being assassinated by rival factions. Even truck drivers shipping oil or moving supplies for the military in areas of unconventional fighting are vulnerable to attack. Regular soldiers are not trained to be bodyguards and are typically too overstretched to take on this burden. Contractors have responded to this demand, working not to defend entire countries but rather to protect high-value individuals. The objective has been remade from the geographical goal of territorial defense to an individuated goal of protecting lives. Their mission success is judged by whether the principal survives. An individual's life therefore becomes the central tactical consideration and the key to continued profitability.

Once again, the profit motive combines with individuation in dangerous ways. In a war zone, as in the civilian economy, a focus on narrow objectives promotes the externalization of risk. Just as a company might avoid cleaning up industrial waste because its responsibility is to ensure that shareholders profit and not to promote a healthy community, contractors externalize the costs of their violence. They react quickly and with lethal force to potential threats, prioritizing the security of an individual over promoting good relations with people in a contested area or achieving other larger military objectives. Those who defend private military forces rightly point out that there have been instances of them acting benevolently to help others in contested areas, even when they did not have a contractual obligation to do this. Nevertheless, the individuated strategic orientation urges contractors to prioritize defense of their employers over other considerations, even when other moral concerns might contravene this. At best, contractors may be left with a dilemma between conflicting loyalties, such as whether to protect their employer or to assist endangered civilians, and be forced to grapple with the question of how they should act. At worst, the interest in protecting the employer may lead them to act in ways that endanger others, as they arguably did in Iraq and Afghanistan by using excessive force to guard diplomats only to harm civilians and undermine the overall strategic goal of securing popular support.

Conclusion

Private military contractors constitute a major shift in the character of armed conflicts and an overt challenge to the Westphalian system. They contravene the longstanding norm that wars should be waged by patriotic citizens out of a sense of loyalty to their fellow citizens and the state. They are likewise responsible for blurring the boundary between combatants and civilians, as well as for perpetrating many illegal attacks. Nevertheless, contractors perform important roles that lie outside the remit of state forces. They may intervene quickly in humanitarian disasters or protect global commerce. They exist because there is demand that cannot be satisfied by regular militaries. To their credit, states have attempted to address some of the problems associated with contractors by imposing new regulations. And defenders of military privatization have ably shown that many of the moral objections to employing them are misguided. The implications of privatized military force are best appreciated not by speculating about what effects idealized regulation could have or imagining hypothetical scenarios in which disinterested contractors select their wars based on moral considerations, but to instead anticipate their behavior based on what we already know about how market incentives guide private organizations. That is to say, we should expect contractors to pursue war in the same way that other actors in the free market pursue their commercial endeavors: by breaking free from geographical restrictions, increasing leverage over those buying their services, shifting negative externalities onto other actors, and ensuring their own profitability.

Just like mercenaries of the past, who specialized in revolutionary weapons, many contractor firms are pioneers in the use of the weapons technologies I covered in the first part of the book. Some help to identify targets for drone strikes, provide cyberwarfare capacities, and use nonlethal weapons to bring military-style security outside battlefields. They therefore not only mark one of the distinctive changes in twenty-first-century warfare but also participate in the other trends I discuss. In the next chapter I turn to special operations forces, which are not coincidentally where many contractors get their start. Private firms eagerly recruit these specialized warriors and proudly list the elite units they draw members from. These two groups should therefore be seen as part of a continuum of elite warriors crossing between private and state employment, with shared membership and practices breaking the boundaries of the military profession and reducing state control over coercive force.

5 • Emulating the Enemy

Special Operations Forces as Insurgent Analogues

In an era of heavy reliance on advanced weapons that overshadow or obviate the human element, special operations forces preserve a timeless sense of a warrior archetype and protect the existential conception of war. "They still represent human agency, battling against the odds and exercising their own discretion. They are still subjective beings who experience danger by operating behind the lines and by enduring more, both physically and emotionally, than other soldiers."[1] Above all, they indicate that individuals can have a decisive influence on the course of wars. Time and again individual special operators have achieved the status of minor celebrities through careers that would ordinarily consign them to anonymity, including professional football player turned Army Ranger Pat Tillman, Chris Kyle (The American Sniper) Marcus Luttrell (The Lone Survivor), and Matt Bissonnette, who wrote about the mission to kill Osama bin Laden.[2] Even those who do not appear on the news take on a special aura conferred by their membership in elite units. They belong to a secretive world that is a source of endless fascination for the public and is a fixture of entertainment. Despite their secrecy, special operations forces are vital for shaping perceptions of conflict. They are shadowy, mysterious figures, yet they also write dozens of memoirs and receive a disproportionate share of the attention in popular culture. They are the

1. Christopher Coker, *Waging War Without Warriors?* (Boulder: Lynne Rienner, 2002), 82.

2. Chris Kyle, Scott McEwen, and Jim DeFelice, *American Sniper: The Autobiography of the Most Lethal Sniper in US Military History* (New York: William Morrow, 2013); Marcus Luttrell, *Lone Survivor: The Incredible True Story of Navy SEALs Under Siege* (London: Sphere, 2014); Mark Owen and Kevin Maurer, *No Easy Day: The Firsthand Account of the Mission that Killed Osama Bin Laden* (New York: Penguin, 2014).

heroic element of war at a time when existential conceptions of heroism are threatened.

Like contractors, special operations forces embody elements of pre-Westphalian warfare dominated by elite warriors. They are not entirely new but are the re-emergence of something familiar in a novel form. Paradoxically, the same technological advancements in long-range and mediated violence that are removing many soldiers from the battlefield allow special operations forces to preserve face-to-face violence and the decisive clash of arms. Operators demonstrate that technological progress in war does not follow a linear path and that technology is not always associated with growing instrumentalism. Innovations can have mixed effects, depending on how they are utilized. This is yet another reason for focusing on thematic changes in warfare rather than imposing strict typologies that all types of combat must conform to. Technological assistance is vital to giving these soldiers their special powers. They can move faster with the help of dedicated transportation units. New types of body armor and stealth technologies give them more security. Their technological modifications upset the rough equality between combatants that prevailed from the eighteenth century to the twentieth, which was grounded in a shared vulnerability to firearms and other gunpowder weapons, and it partially returns warfare to the more asymmetric violence that typified preindustrial combat.

Operators manifest a sense of rebelliousness and individualism from within military services that have long sought to subjugate those qualities in the pursuit of a conventional war ideal in which all participants are treated as relatively anonymous servants of the state. They are often exempted from uniform standards. Operators have come to look more like renegade motorcycle gangs with their own unique weaponry and equipment. In an era of military standardization that extends from the mass-produced equipment down to identical grooming, operators protect individuation as an operational necessity. In some instances, this even allows them to blend in with civilian populations in much the same ways as private military contractors. Changes in weapons and tactics in turn give way to cultural and normative changes. Special operations forces strive to preserve traditional martial virtues and a warrior ethos built on existential warfare even as armed forces more broadly shift away from these toward instrumental war that is heavily grounded in law enforcement. This in turn drives a wedge between them and the conventional forces, as evidenced by the rise of semiautonomous special operations commands.

Despite their heightened individualism and their unique relationship with technology, special operations forces participate in the same overarching themes that mark changes in military operations more broadly. They depend on increased risk-management capacities yielded by their technological augmentation. In particular, their fighting prowess rests heavily on their ability to gain an asymmetric advantage at the tactical level through physical mediation. For example, their sophisticated night vision technologies mediate the visual experience of fighting in a way that is so profoundly advantageous that it has led to a strong preference for nighttime operations. Special operators likewise reflect the broader mission of relying more heavily on forms of military force that offer greater control over the political risks associated with fighting. They allow states to more readily employ precisely calibrated violence in conflicts that fall short of conventional war, especially counterinsurgency, targeted killings, and degradation. Responding to a variety of contingencies around the world has led operators to define themselves as being supra-geographical, able to operate in any terrain and beyond the boundaries of declared wars.

In talking about special operations forces throughout this chapter, I am referring to elite combat units like the United States' Combat Applications Group (Delta Force) and Naval Special Warfare Development Group (Seal Team 6), the British Special Air Service, Poland's Jednostka Wojskowa GROM, and Russia's Spetsnaz, as well as associated special infantry, air support, and logistical units that facilitate special operations missions. Special operations forces are not new in the same sense as drones or cyberweapons. Elite fighters who dominated the battlefield can be found throughout history, and the major conventional wars of the twentieth century involved many specialized units. Throughout this chapter I highlight important historical parallels and how these provide clues about the political and social significance of elite warriors. Nevertheless, like contractors, the form of special operations forces is changing in important ways because of their participation in the themes that I trace throughout the book. These units have changed considerably in recent decades, especially as their role has expanded to cover virtually every kind of conventional and unconventional conflict. They likewise play a much larger role in ongoing operations, military doctrine, and political calculations.

Military Elitism

For most of history, wars have involved elites who could qualify as special operations forces in the sense of possessing superior skill or performing unique tasks. The political character of these warriors varies considerably.

Some, such as knights and samurais, combined high social status with military expertise, using the latter as a way of guaranteeing the former and limiting competition. Other specialized warriors have been more politically subservient. Janissaries and mamelukes were virtual slaves, raised by the state from a young age as fearsome warriors. Other elite units like Napoleon's Guards and the paratroopers of the Second World War were created with a more egalitarian spirit and commitment to meritocracy. The enthusiasm for warrior elites is therefore a persistent theme of war, while the exact political character of the elites is not. Specialized warriors do not necessarily indicate socioeconomic inequalities or undemocratic governance. However, regardless of their own status, military elites tend to form a key part in the ruling establishment to some extent by helping to protect it, extending its power, and contributing to its legitimacy. That is, they may not personally be powerful, but they are often much closer to political power than regular soldiers. Janissaries and mamelukes were slaves, but served as personal bodyguards for rulers. Napoleon's Guards were supposed to be model republicans, but in practice defended the emperor and were spared the worst hardships of his campaigns.

Control of political power or proximity to it means that elite warriors tend to receive privileged treatment and that they are in turn expected to show greater loyalty. They are imbued with symbolic importance as representatives of the regime's good treatment of personnel and of the selfless sacrifice expected from citizens. Until recently, specialized units wore more elaborate uniforms and were featured prominently in symbolic displays of power, such as parades and public ceremonies. They were essential to presenting a strong face to the world and deterring potential rivals. States now tend to keep their special operations forces in the shadows, but they still benefit from general awareness that these forces exist and that they can be called on in emergencies. Elite warriors also tend to be more heavily involved in preserving the political power structure from domestic opposition than other types of units. Authoritarian states employ them as bodyguards and thugs to intimidate rivals. Democratic states tend to employ them in more defensive roles, such as for hostage rescue. In this way, elite warriors tend to transcend international and domestic operations and cross the line between war and politics.

Special operations forces proliferated rapidly throughout the twentieth century to perform niche roles that regular units were not trained for. In this respect they are akin to contractors, which as I showed in the previous chapter are often attractive because they are able to do work that is beyond the remit of conventional units. Germany created Stormtroopers to break through enemy trenches during the First World War. During the Second

World War, new units formed to fight across different terrain types. Of particular importance was the Allies' support for partisans within occupied areas. Special Operations Executive (SOE) and Office of Strategic Services (OSS) personnel infiltrated enemy territory to stage joint missions with partisans at a time when conventional ground forces were waiting for the invasion of Europe.[3] The Cold War led to increasing special operations involvement in unconventional warfare, counterterrorism, and hostage rescue, as well as specific tasks like locating downed pilots or planting explosives on ships. The world's most elite units, such as the UK's Special Air Service, Israel's Mossad, and the US Delta Force became preoccupied with counterinsurgency and counterterrorism throughout the second half of the twentieth century, which prepared them to fight the War on Terror in the twenty-first century. The overall trend during the twentieth century was therefore the proliferation of elite units designed for expertise in a particular mission type or terrain.

In the twenty-first century special operations forces continue to grow and to take on more prominent roles, but most are growing beyond narrow specializations. Counterterrorism and counterinsurgency have become the defining mission type for a range of special units that were initially designed to perform a more varied array of mission types. This brings these forces away from conventional missions that helped to spawn them and into the liminal zone between war and peacetime law enforcement, where they play a much larger role. Special operations forces in the United States, United Kingdom, Russia, Israel, France, India, and the Philippines are now more commonly involved in raids to kill or capture those whose crimes fit more with terrorism and criminal conduct than with enemy military activities. Many are trained by, and even work alongside, law enforcement and intelligence agencies.[4] They learn investigative procedures and conduct prisoner interrogations. They likewise share responsibility for collecting and analyzing intelligence that may lead to subsequent attacks. In short, the trend in special operations is a collapsing of war, law enforcement, and intelligence collection into a single domain that must be mastered in its entirety to successfully counter insurgents and terrorists. The old bureaucratic system that divided these various functions has been replaced by

3. Michael Foot and Richard Daniell, *SOE: An Outline History of the Special Operations Executive 1940–46* (New York: Random House, 2011); Patrick K. O'Donnell, *Operatives, Spies, and Saboteurs: The Unknown Story of the Men and Women of World War II's OSS* (New York: Simon & Schuster, 2014).

4. Sean Naylor, *Relentless Strike: The Secret History of Joint Special Operations Command* (New York: St. Martin's Griffin, 2016).

streamlined, decentralized command and control that make special operations forces more agile and more lethal. Even some of the interservice rivalries and interstate rivalries are disappearing because of interservice and international cooperation.[5] Because soldiers cannot be trained to perform every necessary role, they work across organizational boundaries, often forming ad hoc task forces that only exist for short periods.

Special operations forces are taking on greater importance than they have had in the past, both in the sense of being tasked with more strategically vital missions and in the sense of gaining a greater status as political instruments. They are in the process of supplanting conventional forces not in numbers but in the political importance of the missions they undertake and the ideological role they play in sustaining an existential conception of war at a time when it is generally in decline. Around the world armed forces have found that it is more expedient to use small numbers of elite warriors to accomplish politically sensitive missions. This is true regardless of the specific type of conflict. Special operations forces have led the way during major military interventions in Iraq and Afghanistan, participated in civil wars in Syria and Libya, assisted in the annexation of territory in Ukraine, attacked terrorists in dozens of locations throughout the Middle East, Asia, and Africa, and hunted down leaders of terrorist and insurgent organizations in Iraq, Pakistan, and Somalia. These forces are attractive across mission types because they concurrently reduce the financial costs of fighting, avoid casualties that could provoke a backlash, strengthen foreign allies, and present a heroic vision of war as a reference point for national identity. Above all, they are attractive to politicians searching for easy answers to security threats, a feature they share with the other weapons and tactics I discuss throughout the book.

Full-Spectrum Dominance

Until the twentieth century, it was relatively rare for conflicts to take place in remote regions or in difficult terrain. Pitched battles were confined to areas where armies had room to deploy and maneuver.[6] Protracted campaigns gravitated toward agricultural regions that were rich with fodder

5. Paul McCleary, "US SOCOM Seeks to Broaden Ties With Foreign Forces," *Defense News*, December 3, 2012; Linda Robinson, *The Future of US Special Operations Forces* (New York: Council on Foreign Relations, 2013).

6. John Keegan, *A History of Warfare* (New York: Vintage, 1994).

and food needed to sustain armies in the field, and where they could minimize casualties from disease and poor weather. Mountains, dense forests, swamps, jungles, and deserts were rarely sites of major battles. Even in areas where armies could be maintained, only a relatively small amount of the available land was actually contested, usually land along borders between powerful rivals or at key strategic locations. For the most part, battles were fought for a particular piece of contested geography, which meant that fighting happened in and around that space. Naval warfare was more expansive in principle because of the vastness of the oceans, yet the actual fighting was confined to relatively small areas. Battles were usually fought to control specific trade routes or ports. Large navies equipped for combat and carrying additional men for boarding operations could rarely stay at sea for prolonged periods and had to frequently stop in allied or neutral ports where they could find fresh supplies.[7] These factors, along with the challenges of finding enemies and closing for combat in the open ocean, caused most naval battles to take place within a day's journey to land. Territory, both on land and sea, therefore imposed physical boundaries on war.

Many twentieth-century special operations forces arose from a desire to transcend geographies. Mountain, arctic, jungle, urban, and littoral combat specialists became masters of difficult terrains, bringing combat to more diverse locations. They extended the geographical scope of battle. Previously unassailable barriers were overcome or became a source of tactical advantage. One of the most revolutionary shifts of the Second World War was the use of paratroopers and glider troops, which made a dramatic appearance during Germany's invasion of Belgium, when they landed atop heavily defended positions like Fort Eben-Emael to capture them with little resistance. Airborne units played a major role in Germany's capture of Crete, the Allied landings at Normandy, Operation Market Garden, and Operation Varsity.[8] They were able to seize and hold key objectives to pave the way for heavier units by embracing the new aerial dimension of combat.[9] In later decades airmobile units made up of infantry transported by helicopters introduced yet another new dimension to aerial warfare.[10] Deploying from helicopters

---

7. Keegan, *A History of Warfare.*

8. James Sidney Lucas. *Storming Eagles: German Airborne Forces in World War Two* (London: Arms & Armour Press, 1988); James E. Mrazek, *Airborne Combat: The Glider War/Fighting Gliders of WWII* (Mechanicsburg, PA: Stackpole Books, 2011).

9. William H. McRaven, *Spec Ops: Case Studies in Special Operations Warfare: Theory and Practice* (New York: Presidio Press, 1996).

10. Gordon L. Rottman, *Vietnam Airmobile Warfare Tactics* (London: Bloomsbury, 2012).

allowed soldiers to take advantage of speed and surprise while still maintaining their supply lines and an avenue of retreat, two of the limitations that had previously reduced the effectiveness of paratroopers and glider infantry. The elite forces of the twentieth century had a tendency toward geographical specialization, such that each was primarily concerned with operations within a particular type of space.

Special operations forces of the twenty-first century have taken the more ambitious step of declaring that no terrain is off-limits. The US Navy SEALs epitomize this with an acronym meaning Sea, Air, and Land, which affirms the capacity for launching missions anywhere. Even though the SEALs were created to undertake underwater and littoral missions, they recently spent more than a decade working in landlocked Afghanistan.[11] Other special operations forces similarly train for combat in virtually any environment. Selection and training programs read like a catalogue of terrain types, with courses on jungle, mountain, swamp, desert, and arctic warfare, alongside airborne and air assault training. These units have in a sense become specialized generalists. They aspire to dominate every possible space of war, to be geographically unrestricted. They wage war without being confined to a particular contested terrain or to a specific battlespace, preferring an orientation that is global in scope. It is also common for these forces to go one step further by seeking mastery over the "human terrain," by which they mean the ability to influence the perceptions of indigenous civilians.

One of the trends evinced by the creation of independent special operations commands in the United States, France, the United Kingdom, Russia, Canada, and elsewhere is the distancing of these units from the geographical division of conventional forces into air, land, and naval units, or into regional commands encompassing operations in particular parts of the globe.[12] Special operations supersede terrestrial distinctions and bring their capacities across the entire geographical spectrum. The latest stage in the deterritorialization of war by special operations forces is evident in the operations of the US military's Special Operations Command. This is virtually a new branch of the US military that is made up of units from air, ground, and naval services that are each responsible for pushing the limits of combat in their respective terrains. The elite fighting force contains its own air and naval

11. Dick Couch, *Down Range: Navy SEALs in the War on Terrorism* (New York: Three Rivers Press, 2006).

12. USSOCOM plays an important role in providing a model for these organizations in other countries and pressuring rival militaries to adapt to this new approach. For example, see Tor Bukkvoll, "Military Innovation under Authoritarian Government—the Case of Russian Special Operations Forces," *Journal of Strategic Studies* 38, no. 5 (2015): 602–25.

components, enabling soldiers to deploy unexpectedly, carry out attacks, and withdraw before enemies have a chance to prepare themselves or retaliate. Dedicated drone contingents support the Joint Special Operations Command (JSOC) personnel on missions, as do elite Air Force Combat controllers, who can orchestrate supporting aircraft to magnify the fighting power of Navy and Army ground units. JSOC epitomizes the synergy of forces with their own territorial expertise.

Geographical detachment makes it possible for special operations forces to choose when and how they fight. They must be selective, as they are rarely large enough to seize and hold territory against enemy opposition without assistance from conventional units. They prioritize offensive qualities like speed and surprise, and they eschew defensive obligations (though they do advise foreign militaries and conventional units in these roles). Refusing to operate on land except on their own terms gives special operations forces exceptional powers to manage the risks they are exposed to and to minimize casualties. Quick attacks that can come from land, air, or sea make it possible for operators to achieve local superiority over enemies, then to withdraw beyond the enemy's reach before an effective response can be mobilized.[13] Thus, even though special operations forces continue to face their enemies on the battlefield, they tend to manage risks arising from enemy actions through distance and mediation in much the same way as drone strikes. The selective use of violence, followed by movement beyond the reach of enemy countermeasures, gives special operations an asymmetric character. Enemies are able to fight back, but they are typically placed at a severe disadvantage during aggressive surprise attacks by technologically superior operators.

In some ways this fits with Paul Virilio's prediction that one of the characteristic features of contemporary war is the use of speed to overcome space. As he sees it, "the instantaneous speed of transmission of data, as well as the extreme precision of the guidance and navigation of projectiles, will surpass the destructive power of conventional or non-conventional arms."[14] Speed is certainly an integral component of special operations, a key to gaining local superiority over enemies and evading their responses. However, Virilio's account of causation must be reversed. Rather than an increasing tempo causing special operations to leave geographical boundaries, it is the search to surpass land warfare that has driven the pursuit of greater speed. McRaven characterizes special operations as always taking place against a "well-entrenched opponent"; they transform the enemy's territorial attachments

13. McRaven, *Spec Ops*.

14. Paul Virilio, *Desert Screen: War at the Speed of Light* (New York: Continuum, 2002), 2.

into a vulnerability. Speed is a means to this end, arising through the efforts to break through the Western Front and storm the Maginot Line in the twentieth century, then to hunt down terrorists and insurgents inside walled compounds and cave complexes in the twenty-first. Operations continue to speed up as each new generation of special operations forces makes greater use of integrated air, sea, and land assets. This trend is apt to continue as new weapons technologies grant greater freedom from geographical impediments. Advances in stealth helicopter technology, for example, will give elite soldiers still more freedom to maneuver by air. Drones will allow them to replace heavy ground-based weaponry with dedicated close air support.

As central tools in a global war against terrorism, special operations forces not only move across different types of spaces with ease and attack fortified positions unexpectedly but now also transcend state borders that remain insurmountable for those regular military personnel who remain bound by the strictures of conventional war. Throughout the twentieth century, counterterrorism and counterinsurgency were hindered by the imaginary lines between sovereign states. These bracketed state militaries into specific geographical domains, allowing nonstate adversaries who resisted the Westphalian rules to find sanctuary in states that were beyond reach. Contemporary special operations directly threaten territorial integrity, thereby challenging notions of sovereignty, by following adversaries beyond borders. Thus, US JSOC and the British SAS pursue al Qaeda and Islamic State fighters into Pakistan, Somalia, and Libya (among other places), Russian operators assist paramilitaries in Ukraine, and Israeli agents track down suspected terrorists operating from neighboring countries. US operations have grown to such a scale that they have achieved global reach. Former USSOCOM commander William McRaven estimates that the organization has a "persistent presence in more than 70 countries."[15] General Raymond Thomas says, "On a daily basis, we sustain a deployed or forward stationed force of approximately 8,000 across 80-plus countries."[16] More than any of their deviations from conventional military operations, it is this disrespect for borders that constitutes special operations forces' threat to the Westphalian system. To combat terrorists and insurgents who eschew this model of politics and its corresponding form of war, special operators have sought to emulate them.[17] They

15. William H. McRaven, "Preparing Special Operations Forces for the Future," *Journal of International Security Affairs* 23 (Fall 2012): 10.

16. Nick Turse, "American Special Ops Forces Have Deployed to 70 Percent of the World's Countries in 2017," *The Nation*, June 26, 2017, https://www.thenation.com/article/american-special-ops-forces-have-deployed-to-70-percent-of-the-worlds-countries-in-2017/

17. Turse, "American Special Ops Forces."

join their adversaries in pursuing war beyond borders and therefore bring the powerful states that are ostensibly guardians of the Westphalian system further along the road toward undermining the geographical dimensions that are essential to sustaining it.

Superheroes

Veneration of elite soldiers is a persistent theme across militarized societies. The greatest warriors are held in high regard and celebrated as archetypes of virtue that others should aspire to. This theme is so deeply entrenched that warriors from other cultures are routinely borrowed for additional insight into the warrior mindset, often needing little adjustment or reinterpretation. Classical figures like Achilles, Hector, Cincinnatus, Coriolanus, Beowulf, and Gilgamesh, just to name a few, link living warriors to a mythical past and serve as models of conduct, even featuring as central figures in military ethics training programs.[18] It is a testament to their endurance that these names are probably more recognizable than those of distinguished politicians, scientists, and artists from the distant past. Part of what makes great historical warriors so attractive is that we know who they are. They stand out among the countless anonymous participants in war with disproportionate contributions to victory that warrant special praise. Achilles famously agreed to die young as the price for the glory of being remembered eternally; countless other warriors made an implicit bargain with the same effect upon deciding to become warrior elites and dedicate their lives to fighting.

Ehrenreich says of the premodern elite warrior that "He must have even looked to his fellow humans like a member of some far better endowed species. His horse gave him superhuman height, his armor acted as a glittering exoskeleton from which bladed weapons protruded like tusks or claws, while his heraldry might announce his kinship with lions, leopards, eagles, or other threatening animals."[19] This is an apt description, as elites took steps to cultivate an imposing image with conspicuous displays of military prowess. Tournaments gave warriors opportunities to display fighting abilities. They carried weapons with their own names and lineages and wore armor crafted by

18. Shannon E. French, *The Code of the Warrior: Exploring Warrior Values Past and Present* (New York: Rowman & Littlefield, 2003).

19. Barbara Ehrenreich, *Blood Rites: Origins and History of the Passions of War* (New York: Henry Holt, 1997), 150.

renowned artisans. Such conspicuous displays of skill and status were central to creating a highly individualistic culture of war before the modern era—a culture in which elites participated in combat as commanders of their own retinues and with a high level of personal control over how and when they would fight. That is to say, elite warriors strived for individuation. They appeared as individuals even as they participated in a collective war effort.

Conventional warfare is deeply threatening to individuality, both in terms of how soldiers are organized and in the kinds of violence they are subjected to. Throughout the nineteenth and twentieth centuries, states aspired to the total mobilization of their populations and resources into the war effort. They sought to transform themselves into enormous war machines made up of people who were treated as largely anonymous interlocking parts. Human lives were statistics in the calculus of conventional warfare. Production numbers and supply schedules took on greater relative weight with the growing power of weapons, overshadowing the human element of combat. Powerful new weapons devalued the qualitative differences between individuals because there was little reason to invest heavily in expensive specialized units that would be torn apart by machine guns and artillery barrages almost as quickly as regular forces. The aim of conventional war was inherently contrary to individualism, as it was to inflict as many casualties as possible on the enemy, sometimes even trading lives outright in attritional combat to overpower smaller adversaries through sheer numbers. Soldiers were subjected to artillery, air strikes, poison gas, and machine guns that often seemed profoundly unfair in the apparent randomness of their destruction. Military cultures maintained a high status for their great heroes even as the conditions of fighting and opportunities for individual distinction changed, but it was impossible for individuals to have the same impact unless they were armed with the immensely destructive new weapons. Thus soldiers lost their individuality through the organizational demands of integration and the mode of fighting itself, and they could only reclaim it through technological augmentation. During the nineteenth and twentieth centuries, the Unknown Soldier that represented anonymous conscripts supplanted the warrior elite in the national consciousness.

The Overland Campaign during the American Civil War, the Franco-Prussian War, the Russo-Japanese War, and the First World War epitomized conventional warfare at its most dehumanizing, with soldiers being sent into suicidal frontal attacks in which they were at the mercy of machine guns and artillery, new forces of destruction that had become the centerpieces of war.[20]

20. Geoffrey Wawro, *The Franco-Prussian War: The German Conquest of France in 1870–1871* (New York: Cambridge University Press, 2005).

Many commanders were less concerned with maneuver and position than with simply mobilizing the greatest volume of destructive power.[21] Novels and memoirs from veterans, such as *All Quiet on the Western Front* and *Storm of Steel* emphasize the apparent randomness of death and the feeling of being lost amid monumentally destructive forces. Rapid advancement in killing power took a heavy toll on soldiers and further urged commanders toward massed tactics based on the pursuit of victory through numbers. Such combat was antagonistic to the idea of qualitatively superior martial skill, and it undermined the sense of individuation that had been the hallmark of specialized warriors. It also promoted the kind of figurative and literal entrenchment that special operations forces would later exploit.

The resurgent interest in specialized warriors from the end of the First World War onward marked a partial return to traditional sensibilities of valuing the human element of war. One of the distinctive features emerging during the Cold War, and becoming more pronounced during the War on Terror, is a profound new respect for specialized individuals within the context of politically sensitive counterterrorism and counterinsurgency operations. When it comes to hostage rescue missions in which anything less than perfection results in failure, targeted killings in undeclared war zones, or collaborations with foreign militants and security forces, quality matters far more than quantity. In these cases, states cannot bring their entire arsenals to bear without jeopardizing their strategic and political objectives. The overwhelming force of conventional war compromises legitimacy and invites retaliation. Moreover, enemies lacking the means of fighting force-on-force attritional battles may be driven underground. Special operations forces aspire to use violence that is perfectly calibrated to sensitive political contexts. The incentives of war have once again made elites indispensable. Their recurrent participation in these difficult, high-profile missions has in turn helped to reawaken some sense that individuals still matter in war.

Individuation is a theme across the various issues I discuss throughout this book, but it takes on a distinctive character when it comes to special operations forces. Here it is not simply the increased salience of individual participants in war but also a method of fighting that highlights the continued relevance of people when war is heavily defined by machines. Dick Couch points out that "Amid all the transformation and innovation brought on by technology and global military requirements, we still cling to the tra-

21. John Ellis, *Brute Force: Allied Strategy and Tactics in the Second World War* (London: Andre Deutsch, 1990); William Philpott, *Attrition: Fighting the First World War* (London: Little, Brown, 2014).

ditional, sometimes ancient, and even noble notions of what a warrior is and what we expect of warrior conduct on the battlefield."[22] It is this longing for an imagined past era of war that makes special operations forces so popular and seems to make them profoundly more attractive to the public than drones, even though they are involved in the same kinds of covert missions. As Coker says, "We honor these warriors because we still acknowledge that they are the last link with our own past. They put us in touch with the power war once had to define our humanity, one it now seems to lack."[23] Contemporary special operators reflect a resurgent individualism in war, a sense that human combatants can once again individuate themselves through a combination of martial skill and superior weaponry. And they reassert this individualism in a way that is far more familiar and comfortable than when it comes from computer technicians designing a worm or a drone operator launching a missile.

This individuation through these elite warriors is not simply a revival of traditional sensibilities. It takes on a new form because special operations forces are highly secretive and engage in covert missions. At times the missions are classified and carried out with little public awareness beyond speculative attribution in newspaper reports. In other instances, missions are acknowledged, but those involved have their identities hidden. Photographs of operators in the field are routinely censored, with faces blurred and unit badges covered. They cannot be immortalized by name like Achilles. Paradoxically, the veil of secrecy is a core part of what makes it so easy for elite fighters to capture widespread attention when they do enter the public sphere. Shadow warriors and covert missions are endlessly fascinating. Secrecy heightens the symbolic importance of these personnel and their distinctiveness as individuals or small groups. Efforts to maintain secrecy create an allure that guarantees notoriety whenever identities are revealed. The sense that we do not know exactly who is behind the covert attacks means that membership in one of the elite units is usually an important distinction in itself, especially when a particular unit is credited with carrying out an important operation (such as Seal Team 6 being linked to bin Laden's assassination). Each individual can stand out, and stand apart from the rest of the military, through their membership. Many units claim a kind of ahistorical existence, such that they will always exist and always keep the memory of past warriors alive.

---

22. Dick Couch, *A Tactical Ethic: Moral Conduct in the Insurgent Battlespace* (Annapolis, MD: Naval Institute Press, 2011), 21–22.

23. Coker, *Waging War without Warriors?*, 82.

Special operations forces strive to maintain the image that they are "quiet professionals" who avoid attention. However, this image is doubtful considering how many members of elite units stray into the public sphere. Two of those who took part in the bin Laden mission wrote books about the raid and other missions.[24] The founder of their unit, Richard Marcinko, not only wrote memoirs but also went on to produce fictional accounts of its operations. Dozens of other members of elite units have publicized their experiences, giving them a disproportionate voice in constructing popular narratives about recent wars. Mark Urban says of the British operators that "many of those on the inside allow the ego that comes with their 'special' status or superb physical fitness to evolve into a desire for publicity."[25] Elite warriors likewise work as consultants for military films, and sometimes even as actors.[26] Operators appear in countless films, novels, and video games, thereby promising the kind of immortality that Achilles once traded for his life. Many of these figures are fictional, but they have enormous symbolic value as heroic figures. Even fictional figures help to advance the larger cult of the elite warrior. After all, we know Achilles only as a character in a story, and he is the pre-eminent warrior archetype of Western civilization.

Popular media represent these soldiers as exceptional warriors who often seem to be nearly immortal. They rarely need to reflect on the merits of fighting or consider their own mortality, as they are usually too busy raiding enemy bases or assassinating terrorists. Perhaps most importantly of all, the elite warriors are shown having a direct and decisive effect on the outcome of wars. Their missions are not part of a collective effort by a huge mass of soldiers but rather the work of a small group that succeeds in having a disproportionate influence on the course of events, assassinating enemy leaders, blowing up key targets, or rescuing hostages. This narrative reinforces the prestige of special operations forces. It contributes to the individuation of these soldiers and affirms their status as exceptions to the normal rules of war who can always be trusted with this immense responsibility because of their superhuman abilities.

---

24. Owen and Maurer, *No Easy Day*; Robert O'Neill, *The Operator: Firing the Shots that Killed Osama Bin Laden and My Years as a SEAL Team Warrior* (New York: Scribner, 2017).

25. Mark Urban, *Task Force Black: The Explosive True Story of the Secret Special Forces War in Iraq* (New York: St. Martin's Press, 2010), xiv.

26. John Anderson, "On Active Duty for the Movies (Real Ammo)," *New York Times*, February 12, 2012, https://www.nytimes.com/2012/02/19/movies/act-of-valor-film-with-active-duty-members-of-navy-seals.html

Standing Out

Distinguishing between regular and specialized forces drives a wedge between them and lays the foundations for rivalries and disputes about the nature of the military profession itself. At times, the higher status of elite units provokes resentment from the countless others in the ranks who are not recognized as elites and whose skill is diminished by unfavorable comparison. The resentment is apt to be especially high as the benefits of elite unit membership increase, when nonelite units show similar prowess in combat, or when elites are insulated from hardships affecting other units. Christopher Spearin notes that among the many sources of disagreement between conventional and special operations forces are "differing opinions regarding how military organizations should look and act."[27] They embody distinct conceptions of the military profession and the norms that should govern it, sowing further confusion about what it means to be part of that profession.

The individualism of elite units conflicts with a broader conventional military culture struggling to enforce a sense of collectivity. Special operations forces generally enjoy greater prestige than members of other military units, receive more glamorous missions, and claim a disproportionate share of resources. The above examples of operators' fame outside the military contribute to this. These soldiers also benefit from professional perks that facilitate individuation and deepen separation from conventional forces. For example, they can customize their weapons and gear in a way that recalls knights and samurais wielding unique weapons with unique names and lineages. Matt Bissonnette, who achieved notoriety for writing about his involvement in the mission to kill Osama bin Laden, said that "none of my guns were standard issue. We all had individual modifications on the trigger and grips."[28] He goes on to describe many inimitable pieces of equipment available only to members of his unit. Customization is a mark of deviance from the strictures of conventional military discipline, a reaffirmation of the personalized weapon of the past over the mass-produced, interchangeable weapons of industrialization. It is one of the many status symbols that not only helps to affirm a sense of individualism but that also sets special operations forces apart from their conventional counterparts.

Another persistent complaint against military elites is that they take

---

27. Christopher Spearin, "Special Operations Forces a Strategic Resource: Public and Private Divides," *Parameters* (2006): 58–70, 58.

28. Owen and Maurer, *No Easy Day*, 45.

competent soldiers away from other occupations, and especially from the important junior leadership roles where they could exert a positive influence on their fellow soldiers. Each time a person is recruited into a special unit, it is strengthened at the expense of conventional forces. The concentration of skilled soldiers in a single unit may arguably leave a military weaker overall than if these soldiers were distributed to ensure a basic level of expertise and professionalism throughout the military.[29] As special operations forces become more distinct from their conventional counterparts, this transfer takes on a parasitic character that leaves regular militaries drained of their most qualified members. The transfer of personnel from regular positions into special operations forces also exacerbates the problems associated with contractors. Many contractors are former special operations personnel, and contractor firms give preference to those with this type of experience. This means that more of the lucrative contractor work will generally be accessible to special operators than to their conventional peers.[30]

Divergent operational priorities compound the sense of difference. During the wars in Iraq and Afghanistan, soldiers in regular units were given dangerous and unglamorous work that was more akin to policing than the conventional combat many had expected when they joined the military. They guarded checkpoints and conducted patrols that exposed them to attacks by suicide bombers and IEDs. Special operations forces were largely used for kill and capture missions, allowing them to swoop in and assault a target, then quickly exfiltrate. They created a more glamorous type of combat and did so without the burdens of controlling territory. Their attacks frequently came unannounced, even to their comrades in conventional units. And it was those in conventional units who were left to deal with the aftermath of wounded and dead civilians, outrage when a raid hit the wrong home, and strained civil-military relations. Efforts were made to repair these tensions as the conflict went on, yet the persistently high tempo of special operations missions ensures that they continually intrude into the domains of conventional units who are still tasked with maintaining control of territory. Conventional and special operations forces therefore performed much different tasks and followed divergent norms. Moreover, the latter often did so by circumventing the bureaucratic hierarchy that is a core element of conventional military institutions.

Secretive units naturally cannot be controlled as conventional units and

29. Roger Beaumont, *Military Elites* (New York: Bobbs-Merrill, 1974).

30. Kateri Carmola, *Private Security Contractors and New Wars: Risk, Law, and Ethics* (London: Routledge, 2010).

must instead be governed either by informal hierarchies or take orders directly from the top levels of the chain of command. In either case, normal oversight procedures are bypassed. This is disruptive in its own right, as Eliot Cohen points out. "By subverting the chain of command, courting favor with politicians, and distorting perceptions they disturb the professional basis of healthy civil-military relations."[31] The US military's JSOC and its predecessors have repeatedly been involved in operations that were kept from the public eye until they were launched, or that remain secret. Much of the appeal of special operations forces lies in this ability for senior government officials to circumvent chains of command by engaging directly with the units on the ground. This reduces the number of actors involved in a particular decision, minimizing the chance of information leakage. However, this comes at the cost of adding distance between military services and linking operators more closely to political actors, which risks their politicization.

Spearin points out that "there is the possibility of antagonism as SOF are often presented as a panacea or a 'silver bullet.'"[32] Disagreements are therefore not only driven by the different work and material rewards but also by a sense that special operators are more politically valuable. The tactics employed during recent counterinsurgency operations reveals this sense of elites' greater political value. With the US military's creation of USSOCOM, conventional units are at risk of being left out of entire campaigns. Ground operations in Libya, Somalia, Yemen, and Syria have been conducted primarily by special operations forces, just as the initial intervention in Afghanistan was conducted by a small contingent of around three hundred operators. The high profile of special operations missions like the bin Laden raid and the rescue of Captain Phillips from Somali pirates has generated disproportionate credit. As politicians and the general public see special operations forces as the choice of first resort, with greater political expediency, they endorse a new conception of the military profession. The pre-eminence of this more individualistic, secretive, networked style of organization marks a shift in the developmental trajectory of state military forces, or at least a shift in the nature of the forces that states deploy during limited, politically sensitive interventions.

Tension between elite warriors and regular forces is a fairly common his-

31. Eliot A. Cohen, *Commandos and Politicians: Elite Military Units in Modern Democracies* (Cambridge, MA: Harvard Center for International Affairs, 1978), 79.

32. Spearin, "Special Operations Forces a Strategic Resource," 58.

torical phenomenon, but it takes on a unique character in this instance because of the alternative model of fighting that special operators are developing. They are not simply more skilled or treated better. They are also taking on substantively new work that pushes them beyond the boundaries of the military profession. Their involvement in intelligence collection is one aspect of this. Special operators participating in the War on Terror work closely with law enforcement and intelligence agencies and are trained to emulate them. They carry out interrogations and investigate the homes of suspected terrorists as though they were crime scenes.[33] They also advise foreign militaries and work closely with foreign officials when "building partner nations' capacity."[34] This assistance can easily wander into the policy domain, as the operators are military ambassadors in a sense. At times they are even involved in nonmilitary matters, such as advising foreign governments on how to improve law enforcement As government reconstruction and nation building have become central aims of counterinsurgency, operators have in turn become integral to civil affairs and the training of local forces; in this sense they exist as much to assist foreign partners as to fight.[35] In some instances, they are even the predominant form of military diplomacy. Although this role is most closely associated with US and UK forces, others are following the same pattern. Russia's Spetsnatz, for example, helped to train, and then fought alongside, the Syrian military.[36] Once again the boundaries around the military profession are becoming uncertain because of the growing proximity of law enforcement and war. This risks politicizing special operations forces by linking them more closely to the preservation of specific foreign regimes.

Selective Visibility

Like weapons technologies I discussed in the first part of the book, special operations forces have become glamorous instruments for removing the risk from war by either reducing the costs of fighting or hiding those costs.

33. Naylor, *Relentless Strike.*

34. https://www.thenation.com/article/american-special-ops-forces-have-deployed-to-70-percent-of-the-worlds-countries-in-2017/

35. Robinson, *The Future of US Special Operations Forces.*

36. Thomas Gibbons-Neff, "How Russian Special Forces Are Shaping the Fight in Syria," *Washington Post,* March 29, 2016, https://www.washingtonpost.com/news/checkpoint/wp/2016/03/29/how-russian-special-forces-are-shaping-the-fight-in-syria/?utm_term=.fd1b0d0969d9

In an age of media saturation and persistent fears that civilian publics will not tolerate casualties, leaders are under unprecedented pressure to accomplish objectives that seemed to require the use of military force while also facing stronger pressures from the media, domestic public opinion, and international organizations. This encourages them to rely on special operations forces that hold the promise of accomplishing missions without being noticed, or that at least incur lower costs, which might not provoke the same level of public outrage.

Special operations forces around the world continue to be highly secretive, in some cases not even being acknowledged by their own governments. They are best understood not as purely covert forces but rather as flexible tools of perception management. As I have argued, operators are not always hidden. They are characterized by *selective visibility*. They are simultaneously secretive and intensely public. They are disavowed, yet take a prominent role in political and popular culture narratives. Covert operations are deniable and may make use of capacities unknown to the enemy, but the act of keeping capacities hidden also works to promote hopes (or fears) about what these capacities are. Special operators and their triumphs must be publicized whenever this is beneficial, especially when key moments like the assassination of Osama bin Laden can be used to reinforce the narrative of their omnipotence. Even the Russian government, which is circumspect when it comes to military affairs, has publicized its operators' efforts to destroy Islamic State.[37] At the same time, more sensitive missions and failures must be hidden as much as possible to avoid incurring public opinion costs. The net effect is a narrative of perfect efficacy in which adverse consequences are hidden and benefits promoted to deter adversaries and encourage friendly audiences.

The foremost benefit for receptive audiences is reassurance. Cohen perceptively notes that the tendency of seeking an easy solution to serious crises creates pressure to rely more heavily on elite warriors. "Elite unit prominence occurs only during a politico-military crisis, for it is then that the public searches for heroes and politicians look for panaceas."[38] It should come as no surprise that special operations forces have become a core part of the War on Terror. Since 9/11, terrorism has been depicted as a uniquely dangerous threat acting on a global scale to produce a war with no foreseeable end.[39] The wars

37. Gibbons-Neff, "How Russian Special Forces Are Shaping the Fight in Syria."

38. Cohen, *Commandos and Politicians*, 95–96.

39. John Mueller, *Overblown: How Politicians and the Terrorism Industry Inflate National Security Threats, and Why We Believe Them* (New York: Simon & Schuster, 2009).

in Iraq and Afghanistan failed to provide victories that could allay public fears and restore certainty. Special operations forces provide a reassuring sense that a solution can be found in emulating the adversary's penchant for unconventional warfare. Moreover, they offer many small victories around the world. In lieu of a decisive victory in conventional combat or a successful regime change, rescued hostages and assassinated enemies give some sense that wars are moving in the right direction and accomplishing something.

Western states face a recurrent critique that leaders and their populations are risk-averse, that they lack the will to fight. Osama bin Laden and Saddam Hussein claimed that the United States and other Western powers would not put soldiers at risk or be able to tolerate watching them arriving home in body bags.[40] This sense that the world's most powerful armed forces can be beaten by much weaker, though highly motivated, adversaries undermines those armed forces' deterrence credibility. This gives weak states and non-state actors greater self-assurance when acting against international pressure. Special operators answer this critique with a statement of pure will. As super-human combatants who still represent an existential sense of combat, they are sources of profound national pride because they tacitly affirm that wealthy countries have not lost their fighting spirit. They may also offer a more credible deterrent than cruise missiles and drones can. Those weapons are sometimes critiqued as embodying a Western aversion to war, perhaps even as symbols of cowardice.[41] Operators counter that narrative by continuing the tradition of face-to-face combat.

Of course, covert special operations missions have much in common with drones and cyberattacks, especially when it comes to mediation. They make it possible to fight enemies using lethal strikes that are rarely visible to those who lack direct access to events. It is this type of mediation that makes it possible to manipulate information about special operators and their missions. Control over mediation permits selective visibility. When it comes to covert missions, there are no war correspondents, or even embedded reporters, to challenge official narratives, and no legions of civilians with smart phones recording attacks. The information chain between events on the ground and audiences is almost entirely severed, leaving only the strands that pass through government control or occasionally information pieced together by

40. Christopher Gelpi, Peter D. Feaver, and Jason Reifler, *Paying the Human Costs of War: American Public Opinion and Casualties in Military Conflicts* (Princeton: Princeton University Press, 2009).

41. Fred Aja Agwu, *Armed Drones and Globalization in the Asymmetric War on Terror: Challenges for the Law of Armed Conflict and Global Political Economy* (London: Routledge, 2017).

investigative journalists struggling to corroborate the official narrative after an attack has taken place.

The result of this form of mediation is to increase the power of governments over information about conflicts, positioning themselves as the only real source of news about strikes. This in turn empowers governments to present the strikes as they wish, with little risk of their stories being seriously challenged by outside sources. This is mediation as information control, or at least potential information control. It is generally impossible to even say that the information governments release about covert missions is inaccurate. When governments report all the details with minimal bias, we rarely know it. Without independent sources, audiences are left with profound uncertainty about the nature, extent, and effects of covert wars. This explains the paradoxical visibility and invisibility of special operations forces. Their appearances and disappearances are carefully calculated by policymakers and members of the units themselves to cultivate the right image. Thus we hear about the missions to kill Osama bin Laden, al Zawahiri, and the captors of Captain Phillips, but not about the countless other missions that happen every night in contested areas.

Selective visibility also benefits special operators themselves, helping to advance their own interest in continuing to exist in an era when the human element of war is under threat. Ehrenreich argues that "The point of war is to win, but elite warriors have another aim too, which is to preserve their status as an elite."[42] This is the logic of warrior elites throughout history, to dominate the battlefield and to do so in a way that reinforces their position at the top of a social hierarchy. In the past this goal was often linked directly to the military elites' own political power or indirectly to warriors' service as guards for important officials. Ehrenreich's statement remains true today, but the status being protected has narrowed. The operators themselves are far removed from direct access to political power or even indirect influence over it through their ability to coerce elites (at least in the strong states I am focusing on). They are therefore left to protect their own status, to protect the last vestige of existential warfare as it comes to terms with the technological advancements that threaten it, in an effort to reimagine traditional combat in a way that will preserve its relevance. Selective public engagement makes it possible to simultaneously achieve both of the goals Ehrenreich points out. Victory and the preservation of status hinge on effective perception management. Operators must prevail in high-profile missions because the objectives themselves matter and because the operators have had to struggle against the instrumental conception of war that privileges technological advancement.

42. Ehrenreich, *Blood Rites*, 179.

Mediating the Human Element

Special operators often face their enemies directly; they parachute or take helicopters into hostile areas to engage in close quarters battle. For the operators, the experience of battle is frequently the same kind of direct engagement with enemies that war has historically been. It preserves the feel of authenticity that is disappearing across a broad range of other military specializations that are increasingly detached from anything soldiers of past generations would have recognized as combat. War likewise remains very real for the targets, whether they are terrorists, insurgents, uniformed personnel, or civilians. Nevertheless, special operations forces tend to rely heavily on forms of mediation that protect them from some of the most hazardous assignments.

First, contemporary operators depend heavily on technologies of physical mediation for support even when they are directly engaged with enemies. These warriors operate under extraordinarily difficult conditions. Far away from allied units, and sometimes in countries where there is no official war, they cannot easily call on assistance. Difficulties of allocating support at long distances, especially when it comes to fire support and intelligence, has traditionally limited the utility of special operations forces and rendered them vulnerable when deployed away from supporting conventional units. These challenges are alleviated by pairing operators with unmanned aerial vehicles (UAVs). Because they can conduct strikes virtually anywhere and provide real-time information about enemy movements, UAVs are vital for ensuring that operators can act with relative autonomy. The aircraft can provide real-time surveillance of enemy positions and launch attacks, all without the need for aircraft with onboard pilots. Here the development of mediated warfare in other domains has been central to the rise of a new mode of covert violence. It has also added to governmental control over information by ensuring that involvement in these operations is kept to a minimum.

Second, special operations forces rely heavily on technological augmentation to make ground combat asymmetric. The world's leading special operators are adept at using night vision—a form of physically mediated vision—to gain a decisive advantage over poorly equipped terrorists and insurgents. Throughout history, fighting at night has offered potential advantages in terms of surprise, but has usually been considered too dangerous because of the challenges of maintaining order and avoiding fratricide. Night vision alleviates these problems by transforming soldiers' perceptions while leaving enemies enmeshed in those perennial challenges. JSOC's aviation wing, the 160th Special Operations Aviation Regiment, earned the nickname "Night Stalkers" because of its specialization in nocturnal operations. This specialization was essential because America's pre-eminent covert warriors have a strong preference for night fighting. A major concern for the British Special Air Service during its campaigns in Iraq and Afghanistan was securing night-vision equipment that would allow them to undertake the same nocturnal orientation as their American counterparts.[43]

Third, special operations forces create an institutional mediation of conflict by acting through or alongside local auxiliaries. One of the most interesting efforts to mediate special operations comes from the US Army Special Forces, also known as the Green Berets, which were developed with the goal of training, then cooperating with foreign guerrillas, helping them to launch insurrections against oppressive governments. Although the Special Forces remain more heavily committed to this form of proxy warfare than other elite units, the use of proxies has become widespread throughout US, British, and Russian special operations. The invasion of Afghanistan in 2001 was accomplished with an initial contingent of around three hundred operators drawn from across the US military and intelligence services who were able to work alongside members of the Northern Alliance and other militias opposed to the Taliban.[44] British counterinsurgency doctrine has long relied heavily on local auxiliaries to make up for manpower shortages, with elite units often taking a central role in training fighters then leading them in combat. Russian special operations forces cooperate with Ukrainian militias and the Syrian army in much the same way, first providing training, then fighting alongside those units.

In the United States, cooperation with local fighters through programs like Security Force Assistance (SFA), Security Cooperation, and Joint Combined Exchange for Training (JCET), has become one of USSOCOM's defining missions. They work throughout countries where Americans are actively fighting and in countries attempting to build counterinsurgency capacities, such as the Philippines and Colombia. In these cases American forces are even further removed from kinetic operations and more heavily entrenched in an advisor role through which they can build local capacities in ways that advance US interests without endangering US soldiers or giving the appearance of overt participation in hostilities. McRaven argues that "The U.S. cannot address the challenges of tomorrow alone. It will require a

43. Urban, *Task Force Black*.

44. Eric Micheletti, *Special Forces: War Against Terrorism in Afghanistan, 2001–2003* (Paris: Histoire & Collections, 2003); Leigh Neville, *Special Operations Forces in Afghanistan* (London: Bloomsbury, 2011).

global partnership of like-minded entities that can come together to address mutual security concerns," and that "SOF must encourage the growth of long-standing institutions capable of providing freedom of action by their sovereign governments."[45] This is echoed in the Department of Defense's Defense Strategic Guidance, which states that "Whenever possible, we will develop innovative, low-cost, and small-footprint approaches to achieve our security objectives, relying on exercises, rotational presence, and advisory capabilities."[46]

Special Operations Forces and Norms

The elite warriors of twenty-first-century conflicts emulate the unconventional practices of their nonstate actor adversaries, which can lead them into dubious moral and legal territory. Special operators routinely join private military contractors in morally questionable deviations from uniform standards. They grow long beards and hair that would violate the regulations applying to members of regular units. Moreover, they wear a diverse assortment of uniforms and even civilian clothing, which can make them difficult to identify as members of a particular military or even as soldiers at all. Soldiers sometimes resort to using civilian vehicles in an effort to conceal themselves. Members of the British and American special operations forces often traveled this way during their deployments to Iraq and Afghanistan, especially when tracking suspected terrorists in urban areas.[47] Russian operators in Ukraine and Syria have repeatedly acted without visible signs of nationality, allowing President Vladimir Putin to deny their role.[48] Some decisions to depart from the uniform standards that apply to other members of the profession are tactically sound, as they make it easier for small numbers of elite soldiers to avoid attention. Nevertheless, taking these steps breaches military standards, once again disrupting the profession, and, even more seriously, violates the spirit of international humanitarian law. IHL requires soldiers to clearly identify themselves as such by wearing distinc-

45. McRaven, *Spec Ops*, 10.

46. Department of Defense, *Sustaining U.S. Global Leadership: Priorities for 21st Century Defense* (2012), 4, http://archive.defense.gov/news/Defense_Strategic_Guidance.pdf

47. Urban, *Task Force Black*.

48. Bill Chappell, "Putin Says Those Aren't Russian Forces in Crimea," NPR, March 4, 2014, https://www.npr.org/sections/thetwo-way/2014/03/04/285653335/putin-says-those-arent-russian-forces-in-crimea

tive clothing and carrying arms openly. By failing to do this, at least under some circumstances, operators act much like insurgents and participate in eroding the normative constraints on how wars are fought.

The relaxed uniform standards illustrates an interesting double standard in the treatment of violent nonstate actors. States, the soldiers operating on their behalf, and a broad range of civil society actors, such as media commentators and academics, condemn irregular fighters for refusing to mark themselves as combatants: for wearing civilian clothing, operating from inside civilian structures, and opportunistically merging with the civilian population to escape attack. These are fair criticisms, as irregular fighters do often breach international law and put civilians at risk. Nevertheless, emulating these practices brings states into closer alignment with violent nonstate actors. Both participate in effacing key visual indicators of soldier/civilian status, placing civilians at greater risk of being mistakenly attacked.

The unique standards for operators go deeper than those applying to physical appearance: they include a range of permissions to take exceptional actions. Although the extent of their activities is a carefully guarded secret, reports indicate that America's elite soldiers were involved in the "enhanced interrogations" of suspected terrorists, which amounted to torture.[49] Across elite forces, distinctive rules of conduct are the norm.[50] Cohen warns that "elite units may acquire a sense of being a law unto themselves—thinking that their elite status permits them to use extraordinary measures."[51] It is, after all, implicit in the notion of specialized or elite personnel that they may have to operate beyond the boundaries of conventional practices. This presents a threat to formal and informal norms of war.

Regardless of how they behave in practice, the fact that elite units are subject to selective visibility is in itself normatively questionable. Democratic armed forces are supposed to be accountable to military chains of command as well as oversight by civilian government officials. By circumventing the regular command structure, these forces avoid many of the layers of oversight faced by their peers in regular units. Their activities likewise receive minimal oversight from civilian policymakers beyond the executive and intelligence agencies, or from the general public. With minimal knowledge of how wars are being conducted, the public has limited power to constrain use-of-force decisions or to reach informed judgments about how military force is being

49. Alfred McCoy, *A Question of Torture: CIA Interrogation, from the Cold War to the War on Terror* (New York: Metropolitan Books, 2007).

50. Jeremy Scahill, *Dirty Wars: The World Is a Battlefield* (New York: Perseus, 2013).

51. Cohen, *Commandos and Politicians*, 76.

used. One criterion that is often included among just war considerations is a requirement for war to be declared publicly.[52] Declarations help to ensure that belligerents have opportunities to avert war and that they participate in hostilities as part of a collective agreement to fight. Special operations are one of the central methods of circumventing this norm in twenty-first-century conflicts. Because missions are often covert and take place outside of conventional wars, they undermine the honest, public declaration of intentions. This could in turn limit opportunities to seek nonviolent conflict resolution through diplomatic channels.

Conclusion

In an age when war is dominated by machines, special operations forces help to ensure that a human element remains. They represent a resurgence of elite sensibilities about war, with highly trained fighters being distinguished from the masses of largely anonymous conventional "boots on the ground" and given the most glamorous missions. It should come as no surprise, then, that special operations forces dominate news and popular fiction about the War on Terror far more than regular soldiers or even many of the technologically advanced weapons systems. They preserve the romanticism of existential war and perceptions of combat as being a test of skill between adversaries, yet while also giving policymakers and military commanders many of the same risk-management and mediation strategies that were evident with other weapons systems. Moreover, their covert nature makes it possible to hide missions from public view and preserve the option of publicizing events when these bolster support for an administration or a particular conflict.

Special operations create the impression that the human element of war still matters, but they do this while pushing regular ground forces out of combat assignments. They reflect a division of labor that preserves combat abilities and allows many soldiers to be repurposed for counterinsurgency and humanitarian missions. This contributes to the ongoing identity crisis of modern armed forces that are caught between preparations for combat and preparations for policing actions by relegating soldiers in conventional units to the latter role. This disruption of the military profession and its internal norms relating to the use of force is evident from how special operations forces are developing into their own branches of the military while also

52. Brian Orend, *The Morality of War* (Orchard Park, NY: Broadview Press, 2006).

shortcutting the normal military chains of command and creating cross-national links with other elite units. The dominance of elite units seems likely to continue into the future, with the world's leading militaries investing heavily in them and expanding their capacities. These armed forces will gain the power to more readily use limited coercive force that falls short of war, often without even giving the appearance of military involvement.

Part III • Strategic Innovations

6 • Naming Enemies

Targeted Killing and Punitive Violence

Deliberately targeting enemy leaders makes intuitive sense as a military and political strategy. Eliminating leaders shatters the enemy's command structure and may cast those in the lower ranks adrift. It can strike a crushing blow to morale because of leaders' symbolic importance and the trust subordinates place in them. A resulting power vacuum can cause internecine fighting as others rush to fill the void, further destabilizing the enemy. In pre-Westphalian wars, it was common for armies to collapse and flee when their leader was killed or retreated, especially when leaders were fighting for largely personal reasons, as in wars of succession. Throughout the Westphalian era, during which our conception of what qualifies as conventional warfare took shape, deliberately targeting political or military leaders was usually considered illegitimate. Despite its potential effectiveness, there was a powerful norm against killing specific individuals on the battlefield or away from it. This followed from attempts to formalize conflict as an extension of politics divorced from personal motives that would warrant selective attacks against specific people. It likewise reflected the value placed on officers as managers of armed force to maintain order. There was fear that leaderless soldiers might transform into armed mobs.

Targeted killing regained prominence during the Cold War as both sides sought to strike the other in small ways that could avert major wars. Decapitation was seen as a potential substitute for war, or at least a way of using force while minimizing exposure. Moreover, it was hoped that a handful of powerful elites were sufficient to determine a country's regime type and foreign policy, which meant that the international balance of power could be rearranged by eliminating or protecting key individuals. United States intelligence agents attempted to kill Fidel Castro, Ngo Dinh Diem, Patrice Lumumba, and Rafael Trujillo, among others. The Soviet Union assassinated former leaders like Leon Trotsky and made attempts on foreign heads of state, including Josip Broz Tito. States also launched killings beyond their own borders. Soviet agents targeted numerous émigrés and defectors. It was also during this time that targeted killing became a favored approach to counterterrorism. Israel's Operation Wrath of God sent Mossad to kill suspected members of the group Black September who were involved in the 1972 Munich Olympic massacre. Israel's covert strikes killed targets throughout the Middle East and Europe with minimal risk of retaliation. Decapitation has been popular since the end of the Cold War. During the Gulf War, coalition forces followed an American plan of heavily targeting top Iraqi leaders and their communications infrastructure in an attempt to decapitate the regime.[1] Similar efforts were made to track down leading warlords in the Balkans, Africa, and Middle East throughout the small wars of the 1990s.

Contemporary targeted killings owe their popularity to supply side and demand side conditions. On the supply side, new technologies have made it possible to attack enemy leaders more effectively than before. States can go beyond snipers and assassins to utilize drones, precision-guided munitions, and more effective special operations forces. They likewise benefit from advanced surveillance technologies, especially satellites, when locating adversaries. On the demand side, narratives of reducing the costs of violence and prioritizing political over military objectives support efforts to make greater use of small strikes against critical targets. Targeted killings in the twenty-first century have taken on a distinct form because of the methods used, the frequency of attacks, and the role they play in extending violence beyond conventional war contexts.

In past conventional operations, attempts to kill leaders were usually designed to fragment the enemy's forces, facilitating victory against subordinates who lacked direction. Killing leaders was a means of undercutting a collective war effort, rather than the ultimate goal. Contemporary targeted killing is more individuated insofar as it treats enemy leaders as themselves being the real enemies, while casting subordinates as unwitting or unwilling accomplices. Eliminating individual leaders like Saddam Hussein, Osama bin Laden, Abu Musab al-Zarqawi, and Muammar Gaddafi was so important to achieving a perceived victory that targeted killings were imagined as being almost self-sufficient. It was as though merely removing these leaders and a few close associates could cause radical change, even democratize countries. The strategic narrative in counterterrorism and counterinsurgency is that

1. Richard G. Davis, *Decisive Force: Strategic Bombing in the Gulf War* (Air Force History and Museums Program, 1996).

most ordinary combatants are not political ideologues with a stake in the fighting, which means that they can be transformed into allies once leaders are killed.

I define the term "targeted killing" as the attempt to kill a specific enemy who has strategic or political value, either because of who that person is or because a pattern of life makes that person seem threatening. I specify the importance of strategic and political purposes to distinguish this sense of targeted killing from the intent to kill specific enemies for tactical considerations, such as decisions to target officers before enlisted soldiers during a battle. The victim must have great value, or at least be perceived as being valuable. The killing can be of someone who is known by name, which is called a "personality strike" by the US government, or someone who is identified by patterns of life and subject to a "signature strike." Both types of attacks have been a fixture of operations throughout the War on Terror and deserve consideration here. It is with the many unknown victims in mind that I object to Michael Gross' description of these attacks as "named killings."[2] I also avoid any language that could indicate that the targets are always leaders, which is implied by the term "assassination," because the targets are not always in leadership positions. Some of the most prominent targets are propagandists and media figures, which reflects the importance of these attacks for perception management.

Individuals as Enemies

In a sense, targeted attacks were common, even expected, in premodern conflicts because of elite warriors' desire to face their equals. The *Iliad* offers scenes of Greek and Trojan warriors addressing each other on the battlefield and pairing off in single combat. The personal relationship between warriors is prized so highly that each seeks a worthy opponent and fights with an eye to beat that specific adversary just as much as to beat the rival army. Moreover, the personal nature of the conflict means that specific figures take on a disproportionate role in the strategic framework, such that killing one can dramatically tip the balance of the fighting or bring it to an end. Even the motives for the war are grounded in a series of personal disputes, especially over Helen. The conflict feels more like a personal struggle between individuals pursuing their own motives than a political conflict in

2. Michael L. Gross, "Assassination and Targeted Killing: Law Enforcement, Execution or Self-Defence?" *Journal of Applied Philosophy* 23, no. 3 (2006): 323–35.

which collective goals take precedence. This is a fictional example, yet there are plenty of real-world analogues from ancient and medieval wars that show a similar pattern.

Alexander the Great won the Battle of Gaugamela in large measure because of his decision to charge through a gap in Darius' line and head straight for the Persian king. He attacked a specific enemy whom he knew by name, precipitating the Persian army's collapse. Medieval combatants likewise frequently sought out individual opponents both because defeating them could produce a decisive outcome and because this was a route to personal enrichment. Taking a prestigious captive could secure a large ransom. Killing a dynastic rival could bring an end to hostilities. Targeted attacks were encouraged for the personal motives that tended to dictate the overall strategy and the actions of individual participants. This fit with the prevailing norms of war, according to which fighting could be legitimately carried out for private, rather than collective or political, goals. Or perhaps it would be more accurate to say that politics itself was highly personalized because governments were under the control of specific people and often treated as an extension of their households.

The Westphalian system reconceptualized the relationship of European states in ways that laid the foundation for what we now see as conventional interstate warfare. States agreed not to intervene in each other's domestic affairs and were to respect each other as rival sovereign entities that had equally valid rights to rule within their respective territories. This understanding granted each state an unprecedented degree of coherence as a distinctive entity existing apart from the individual leaders. When this international consensus was combined with domestic political reforms, heads of state became the managers of national governments, rather than treating the government as an extension of their personal property. This had the further effect of cutting cross-national kinship and patronage links that had previously served as the rationale for rulers to claim a rival's position, as in the case of the Hundred Years War. Personal wars were replaced by public wars fought on behalf of the nation for reasons of state. Individualized enmity was supplanted by collective enmity. This is exemplified by Clausewitz' political and instrumental understanding of war. As he says, "modern wars are seldom fought without hatred between nations; this serves more or less as a substitute for hatred between individuals."[3] He also argues that "combat in war is not a contest between individuals. It is a whole made up of many parts."[4] The

3. Carl von Clausewitz, *On War* (Oxford: Oxford University Press, 2007), 86.
4. Clausewitz, *On War*, 38.

enemy state, or at least its military, is the proper object of attack, not those exercising control.

Also important was that the Westphalian system gave opposing belligerents grounds for preserving the status of rivals, even those heads of state and senior figures who were responsible for instigating conflicts. The system's stability depended on having leaders in place to guarantee continuity. Relegating the fighting to the men in ranks, while exempting senior leaders and politicians from combat, ensured that wars could be fought to make small adjustments in the international system without triggering fundamental shifts in the nature of political authority. This logic extended to officers in the field. Although it was considered legitimate to kill officers, and they tended to suffer high casualty rates, norms developed to forbid intentionally targeting officers under the belief that they were essential for preserving order. They had to govern the conduct of the men to keep them subordinated to the collective goal, just as senior officers and policymakers had to govern the affairs of state. A low opinion of snipers was one manifestation of this. "The concept of a soldier deliberately stalking a human quarry as one would an animal was to most infantrymen a repugnant one."[5]

Attempts to formalize an international state system coincided with over a century of monarchical consolidation in Europe, during which traditional elites were disempowered and subordinated to institutions. The nature of political authority changed from what Weber characterizes as charismatic and traditional authority to rational-legal authority.[6] Targeted attacks make sense when directed against the former types because these kinds of authorities have personal or dynastic claims to power. They rule because of *who* they are. Rational-legal authority lends itself to bureaucratic institutions, in which leaders lose their value because they are replaceable. This is why targeted killings are rarely used against democracies and republics, in which leaders have limited personal authority beyond what they temporarily gain from an office. Targeted killing is primarily a weapon against nonrepresentative regimes, about which it can be plausibly claimed that specific individuals have a decisive influence.

The preceding examples of intentionally killing enemy leaders before the Westphalian era took place on battlefields. One could argue that killing specific enemies during wars is not really targeted killing in the sense that the term is commonly used now, as a synonym or near synonym for "assassina-

---

5. Martin Pegler, *Out of Nowhere: A History of the Military Sniper* (Oxford: Osprey, 2004), 20.

6. Weber, "Politics as a Vocation."

tion." Killing specific enemies away from the battlefield and with the help of deception has generally been disreputable. Even when targeted attacks during war were legitimate, assassination was not if it involved illicit instruments or deception.[7] Emer de Vattel's definition of assassination as "a murder committed by means of treachery" is indicative of this.[8] Vattel likewise argues that such killings are only appropriate during wartime and when compelled by necessity. This places targeted killings squarely within the normative domain of war. International legal guidelines relating to assassination are murky, but they are generally taken as prohibiting this practice when it happens outside war and is akin to murder. Most legal discussions of assassination focus more on perfidy than on assassination itself. This raises problems when it comes to contemporary targeted killing practices, which often overcome the need for deception and treachery by relying on technological asymmetries.

Domestic accountability mechanisms can also constrain targeted killings. Assassinations outside of wartime were carried out, or at least attempted, throughout the Cold War, but these were almost always covert missions denied by the responsible governments. This constituted tacit acknowledgment that such attacks were not legitimate political instruments. In 1976 President Gerald Ford issued Executive Order 11905, declaring that "no employee of the United States Government shall engage in, or conspire to engage in political assassination." President Ronald Reagan reaffirmed this in almost identical language with Executive Order 12333 in 1981.[9] Even covert attacks against high-profile terrorist leaders were off-limits for American agents during the 1980s.[10] However, in the United States and around the world, the post–Cold War era was marked by a reversal of this norm. With a host of conventional and unconventional operations directed largely against specific hostile leaders, targeted attacks appeared to be morally and strategically preferable to destroying entire armed forces.

Targeted Killing and Norms

Like the other topics I have discussed, targeted killing poses a series of normative paradoxes. It is not more or less moral than alternative methods of

7. Michael N. Schmitt, "State-Sponsored Assassination in International and Domestic Law," *Yale Journal of International Law* 17, no. 2 (1992): 609–85, 614.

8. Schmitt, "State-Sponsored Assassination," 616.

9. https://www.archives.gov/federal-register/codification

10. Robert B. Baer, *The Perfect Kill: 21 Laws for Assassins* (New York: Plume, 2014).

war; it is both at the same time. Targeted killing embodies a desire to perfectly conform to the principles of discrimination and proportionality. It is difficult to imagine greater compliance with these principles than when belligerents restrict the use of force to strikes against specific opponents to achieve objectives without any excessive or misapplied force. Targeted attacks go beyond simply sparing civilians to also protecting enemy fighters who are placed outside the targeting parameters. Proponents of targeted killing emphasize these moral benefits. For example, David says that "targeted killing is discriminatory in that it focuses exclusively on one's adversaries. Civilian casualties and collateral damage are minimized. It is proportionate in that only enough force is used to accomplish the task."[11] Targeted killing thus appears to be the perfect means of simultaneously satisfying operational and moral demands. The same reasoning can be extended to the technologies associated with targeted killing, such as drones and guided munitions. The tools used to conduct targeted killings tend to conform more closely with the principles of discrimination and proportionality than do the weapons of conventional war.

The problem is that conforming to these *jus in bello* principles assumes that there is a state of war in which those principles pertain. When targeted attacks are carried out in a gray area between war and law enforcement, it is unclear whether *jus in bello* provides the correct moral framework. Indeed, the battle lines in the debate over the permissibility of targeted killing are primarily drawn between supporters who see it as a kind of highly restrained warfare and critics who either see it as a kind of degenerate law enforcement or who object to conducting operations in such a morally ambiguous realm. In other words, targeted killing is a morally superior type of warfare, but it may not actually be a type of warfare.

The first normative paradox is that targeted killing's compliance with the principles of discrimination and proportionality is often achieved by diluting the boundaries of war to the point that it is unclear whether norms of war properly apply. In this sense there is a victory for *jus in bello* and a loss for the broader just war project of imposing normative constraints on war. By extension, there is a loss to the organizing logic of the Westphalian system, which presupposes a clear distinction between war and peace. To endorse the kinds of targeted killing that characterize twenty-first-century conflict on the grounds that these are more discriminate and proportionate is to accept that *jus in bello* principles can extend beyond the confines of the war paradigm

---

11. Steven R. David, "Israel's Policy of Targeted Killing," *Ethics & International Affairs* 17, no. 1 (2003): 111–26, 117.

and into law enforcement, or at least into some third category between the two. This transforms norms of war into more general guidelines for use-of-force decisions and tacitly acknowledges that war is not normatively special. Losing a sense of war as a special activity threatens the utility of *jus ad bellum*. That domain of just war theory concerned with when it is permissible to wage war cannot offer a valuable contribution when there is no clear line between wartime and peacetime violence. That is to say, if the same use-of-force decisions are made in war as outside war, and those use-of-force decisions are governed by identical norms, then the imaginary line between war and peace loses moral significance, such that crossing it would not entail the kinds of special considerations that *jus ad bellum* is concerned with. The result is that the means of fighting become a more urgent moral puzzle than the reasons for fighting.

The second paradox is that targeted killing seems to bring too much moralizing into just war theory. Gross correctly points out that the permissibility of targeted killing depends heavily on how enemies are classified. If they are combatants, then they can be killed even if they are not actively fighting or posing a threat, just as enemy soldiers may be attacked when they are eating or sleeping. If enemies are criminals, then attempts must be made to arrest them instead of killing them and it is not permissible to kill them when there is no immediate threat.[12] He argues that the norm against targeted killing during war helps to sustain a sense that soldiers are morally innocent. They are enemies by virtue of their collective membership in a hostile force, not because they have personally done something wrong. This is why military prisoners are entitled to greater respect than criminals, and why they must be released when hostilities are concluded. Because they are morally innocent, enemies are not supposed to be punished. However, efforts to single out specific adversaries suggest that there is personal fault that makes those people liable to attack. The guilt must belong to an individual specifically and must result from actions that person has taken aside from simply being a member of a hostile polity, which would not provide grounds for choosing that person out of the broader group constituting that polity.[13] Thus just war theory attempts to justify the equitable treatment of soldiers by appealing to an innocence from moral guilt at the individual level, while targeted killing relies on the assumption that certain adversaries are guilty of offenses so grievous that they can be attacked as individuals.

Contemporary targeted killings are often imbued with punitive signifi-

12. Gross, "Assassination and Targeted Killing," 106.
13. Gross, "Assassination and Targeted Killing."

cance and carried out with vengeance in mind. The enemies singled out for death are not morally innocent in the way just war theory or the Westphalian conception of conventional interstate war presuppose.[14] Enemies are described as terrorists who are fundamentally threatening, as insurgents undermining the established political order, or as murderous dictators who cannot participate in the international system. They are not on the receiving end of attacks as a matter of political convention. *They are morally abhorrent.* Even when these enemies are captured rather than killed, as in the case of Saddam Hussein, Muammar Gaddafi, and some senior members of al Qaeda, they are denied the rights of enemy combatants and instead treated as criminals who are liable to imprisonment or execution.

The final paradox is that the moral advantages of targeted killing are best realized by engaging in the kind of preventive violence that just war theorists usually oppose. According to just war theory, soldiers on both sides of a war enter into a reciprocal condition of acting in self-defense against the opposing side. Those who bear arms or who assist in the war effort may be attacked because they are threatening, either in the sense of posing a direct threat or in the sense of being essential to the threat posed by the enemy forces as a collective. Wounded combatants and prisoners may not be attacked and are entitled to basic care because they have ceased to be threatening and returned to something akin to civilian status. It is likewise impermissible to launch preventive attacks against the personnel of rival forces when those personnel are not fighting or otherwise contributing to the collective threat. For instance, it is usually not permissible to attack soldiers when they are on leave to visit their families.

Enemies selected for targeted killing are rarely in the midst of some hostile action akin to that of soldiers when they are attacked. This makes it difficult to justify these killings in the name of self-defense. The targets are usually far away from their intended victims, as in the case of American strikes against members of al Qaeda in Yemen on the suspicion that they aim to carry out attacks in the United States. Attacking long before the threats can become real is arguably necessary, but it moves violence into the preventive domain. Just as important is the extent to which people are attacked either for past actions or for actions that would not qualify as participation in war as that concept has traditionally been understood. Many targeted strikes are not conducted to avert an immediate threat but instead to punish past deeds or to manage a future risk. Again this response is understandable given the

14. See Michael Walzer, *Just and Unjust Wars: A Moral Argument with Historical Illustrations* (New York: Basic Books, 1977).

nature of the threat posed by violent nonstate actors; preventive attacks in particular could be morally advantageous if these can be carried out away from innocent bystanders. Nevertheless, attacking opponents in domestic settings and when they are not immediately threatening attenuates the conception of war as self-defense, which is widely treated as the normative framework that renders war legitimate.[15] Here again, the practice satisfies some moral guidelines while posing a fundamental threat to basic assumptions about how war can be justified.

The Search for Vengeance

As the moral controversy indicates, attitudes toward targeted killing at any particular moment do not just reflect on a particular method of killing; they provide insight into how adversaries are perceived. Launching targeted attacks as a form of punishment introduces a normative dimension to war that just war theorists since Grotius and Vattel, writing in the seventeenth and eighteenth centuries, sought to eliminate. They and subsequent just war theorists understood war as a political relationship that should be managed according to shared norms and in which the opposing belligerents had to be treated as legitimate entities with equal rights and responsibilities. Imposing normative boundaries on war depended on first understanding war as a kind of contest between states following the same rules. Even if belligerents were fighting against their will, such as when attacked by an aggressor, there had to be some element of consent to sustain the normative meaning of war as a permissible extension of politics. In other words, the just war tradition's attempt to develop an overarching normative framework for war depended on an underlying assumption that opposing belligerents are morally equivalent when they enter into war and that no one deserved to be targeted as an individual.[16]

Clausewitz reflects the same thinking with his claims that war is essentially a political activity. As a political contest, judgments of good and evil have little place. They threaten to upset the rational calculations that he advocates. Clausewitz goes so far as to compare combat to dueling and wrestling, metaphors that reflect the presupposition of shared rules binding the

15. David Rodin, *War and Self-Defense* (New York: Oxford University Press, 2003).

16. This assumption has been challenged by some recent just war theorists, but it remains the standard view, especially among policymakers and members of the military.

participants.[17] Perceiving attacks as punitive and enemies as evil marks a radical change in the collective understanding of what war is. This undermines the status of war as a contest between equal entities that can assent to the rules and remakes it as a condition that is inherently asymmetric. War goes from a duel to a hunt.

Strange as it may sound, war depends on a certain degree of trust. Conventional warfare can only be conventional if there are shared practices that are explicitly or tacitly agreed to by opposing belligerents. Such norms depend on reciprocity, a formal equality between opposing belligerents that makes it possible for them to in some sense mutually consent to participate in war. Historically, targeted killing was often considered immoral, or at least morally dubious, when this sense of fairness was lacking. Before the twenty-first century, it was largely associated with treacherous tactics, such as dressing combatants up as civilians or feigning surrender. Feelings of fairness continue to shape the moral disagreements related to targeted killing. Gross opposes assassination on the grounds that this violates the trust that is essential to war. For one thing, the victims of assassination do not have the same opportunities to surrender as a soldier would have during a battle. The increased reliance on intelligence operatives in assassinations also undermines trust by encouraging deception and defection. This assault on trust means that "targeted killing provokes a fierce reaction against both collaborators and those who recruit them and, at the same time, rend the moral and social fabric of the community."[18] Despite these concerns about assassination, Gross defends what he calls "named killing." "Targeted killing properly understood as named killing rather than perfidious assassination offers a solution necessitated by the lack of uniforms among combatants."[19] That is to say, Gross manages to justify the practice by removing the kind of deception that makes war seem unfair.

If enemies are perceived as being fundamentally evil and deserving of punishment, then considerations of fairness fail to apply. An immoral enemy has already violated the rules and forfeited the expectation of equal treatment. In many cases, targeted killings manifest a desire for vengeance against enemies who are perceived as fundamentally evil. This gives the attacks a punitive character. Israel pioneered modern targeted killing practices with its campaigns against former Nazis and those who were implicated in Black

17. Clausewitz, *On War*, 13.

18. Michael L. Gross, *Moral Dilemmas of Modern War: Torture, Assassination, and Blackmail in an Age of Asymmetric Conflict* (New York: Cambridge University Press, 2009), 113.

19. Gross, *Moral Dilemmas of Modern War*, 121.

September's attack on the 1972 Olympics in Munich. It pursued these people with a sense of moral purpose. They meant to avenge past wrongs by killing those who were responsible. These actions fell beyond the scope of conventional war, as they were directed against war criminals and terrorists. But this was precisely the problem. The practice of targeted killing emerged and gained momentum in its use against these people, and it became the preferred method of handling subsequent enemies, falling into the gray area between war and law enforcement. The punitive dimension of these missions persisted throughout Israel's attacks on terrorists and insurgent leaders such as Hamas cofounders Abdel Aziz al-Rantissi and Ahmed Yassin. Attacks on insurgents are particularly challenging from a normative perspective, as these people fit more closely within the normative paradigm of war and yet are still denied recognition according to that status.

Killings during the War on Terror have likewise been informed by a search for vengeance. This is clear from the attack on Osama bin Laden. By most accounts, bin Laden was rendered largely irrelevant by the time he was killed. His freedom to act was so constrained that he was ineffective as a leader. Killing him was more a matter of punishment than of preventing future attacks. Even if bin Laden had still been an effective leader, there is little doubt that the decision to launch such a risky attack and the public's jubilant response had as much to do with a desire for revenge than with prevention. Attacks against ideological leaders who appear to have little direct influence in operational matters is further evidence that targeted killings are largely punitive. These people are not directly threatening, such that they might participate in an attack, but they are morally threatening because of how they participate in the radicalization process. The attack on Anwar al Awlaki stands out as an especially clear example. Awlaki was a cleric who had achieved fame first as an advocate for reconciliation between Islam and the West in the days after 9/11, then as an inflammatory imam urging war with the West. Although there was some evidence implicating him in Umar Farouk Abdulmutallab's failed attack, Awlaki was largely a symbolic figure and was also technically protected from attack because of his status as a religious leader. Some supporters of targeted killings even argue that punitive violence is justified.[20]

Targeted killings are not necessarily punitive, but they lend themselves to this outlook. If individual enemies are to be distinguished from the masses, then selection criteria must exist. And once those selection criteria are in place, these are apt to take on a moral dimension. Even if targets are in the first instance selected based on their strategic value or their contribution to

20. David, "Israel's Policy of Targeted Killing."

the enemy war effort, the mere fact that they stand above their comrades and are perceived as more deserving of death invites moralizing. This builds on an existing tendency to demonize enemies to make killing easier, albeit while narrowing the focus of demonization to the individual level. It pushes perceptions of war away from the continuation of politics between equal states and into the domain of law enforcement. It transforms wartime enmity from a political/instrumental relationship between states or groups into political and private enmity imbued with emotional and existential weight.

Authority and Chains of Command

The practice of targeted killing is heavily shaped by the structure of the chain of command on both sides. As I discussed previously, the more centralized an enemy's command structure and the more influential individual leaders are, the more tempting a targeted attack becomes. These attacks are most promising when launched against charismatic or traditional authorities. Such figures are often difficult to replace and tend to centralize control. Representative governments and organizations with complex bureaucracies are poor targets because leaders are relatively easy to replace and have overlapping spheres of authority. Targeted killing is therefore a weapon that is better when used against states and nonstate organizations that are either authoritarian or held together by personal links than when used against bureaucracies. This fits perfectly with the prevailing sense that nondemocratic regimes are fundamentally illegitimate in the post–Cold War world. It is also a natural byproduct of the greater resilience of organizations that distribute power more evenly.

For the attacker, the chain of command matters because it shapes whether and how attacks are conducted. The first important consideration is distance from the target. Soldiers on the ground kill anonymous enemies who they are pitted against as members of a hostile collective; they rarely have opportunities to target enemy leaders or even to attack specific enemies that are known by name. It is difficult to imagine an infantryman on the front lines during World War II engaging in a targeted killing of specific enemies. Defense against immediate threats is a more urgent problem. Commanders who are further from the battlefield are not in danger of physical attack and are instead subject to much different risks associated with maintaining public support and legitimacy, which necessitate more narrowly focused applications of force at the most important targets. That is, personal security via physical mediation offers greater freedom to step away from the necessities of defending against concrete threats and to more carefully select targets based on risk calculations.

The second important consideration is the type of conflict. During conventional wars, it is possible for targeting decisions to work through complex bureaucracies without causing excessive disruptions because violence is generally directed at enemies in the field or at fixed targets. Involvement of multiple actors in the decision-making process helps to manage resources effectively and reduces the chance of mistakes. When attacks are carried out in asymmetric conflicts against mobile and secretive violent nonstate actors who offer limited windows of opportunity, the chain of command must be shortened to the point that senior officials make targeting decisions, sometimes even circumventing midlevel commanders. As Gregory points out, "the late modern kill-chain is increasingly directed at mobile and emergent targets."[21] This is primarily a feature of counterterrorism and counterinsurgency, but it will not be relegated to those domains. As state security forces redefine themselves in terms of asymmetric warfare, they mirror the high mobility, decentralization, and flexibility that make this organizational change so attractive. States will certainly never develop as far in this direction as violent nonstate actors, yet their adaptations build a particular style of fighting that will inevitably carry over into any conventional conflicts they are tasked with.

Greater involvement from senior political officials in battlefield decisions through a shortened chain of command means that targeting decisions tend to flow from high levels of government and the military directly down to drone pilots, special operations forces, missile controllers, and others who are charged with executing the missions. The centralization of control is evident in the way kill lists are managed by top figures in the civilian government.[22] In some ways, this is a noble goal. By circumventing the usual military chain of command, the president takes responsibility for the conduct of the war and made it easier to link decisions back to him. In principle, this might even facilitate efforts to hold policymakers accountable. However, this also marginalizes intermediary ranks, effectively flattening the military hierarchy. Strategic and political decisions are transmitted directly into the tactical domain, and vice versa. The operational levels of military organization are pushed out of the decision-making process. Protracted infatuation with targeted killing has the ironic effect of counteracting bureaucratization and cen-

21. Derek Gregory, "From a View to a Kill: Drones and Late Modern War," *Theory, Culture, & Society* 28, no. 7–8 (2011): 188–215, 196.

22. Chris Woods, *Sudden Justice: The True Costs of America's Secret Drone War* (London: C. Hurst & Company, 2014).

tralizing authority, making the states that carry out these strikes look more like the organizations they are attacking. States do not actually lose their bureaucracies and representative institutions just because of a particular method of fighting. Nevertheless, this style of warfare pushes against these institutions and provides an impetus for centralization. The ironic result, then, is that use of targeted attacks tends to erode the bureaucratic structures that provide resilience against targeted attacks.

The first casualty of centralization is the military profession itself, or at least the conventional military that is not involved in targeted strikes. Conventional forces are kept away from these covert missions, with a limited role in planning them and reduced oversight. To the extent that targeted killings become the dominant approach to fighting a particular adversary, the conventional military may take a back seat in entire campaigns. Preoccupation with targeted attacks has made it possible for the United States, United Kingdom, Russia, Israel, and France to wage long campaigns with minimal mobilization of their armed forces. This produces a strange state of affairs in which wars can be conducted without much military involvement.

Many targeted attacks are not carried out by members of the military but rather by paramilitary organizations, intelligence agencies, or local fighters. These outsiders to the military profession are brought into warfare and given decisive roles that often come to overshadow the work of regular soldiers. Even when military personnel carry out the attacks, they tend to be personnel who fall within one of the paradigmatic modes of twenty-first-century warfare. They are elite special operatives, drone pilots, or missile controllers. What is important here is that some of the most high-profile actions come from atypical military personnel, who then take on disproportionate attention in narratives about ongoing wars. The capture of Saddam Hussein and the assassinations of Zarkawi and Osama bin Laden became synecdoches of the wars in Iraq and Afghanistan for two successive US presidential administrations searching for a victory. The personnel involved in these missions went on to become central figures in media coverage and discussions of war. Other soldiers are reduced to anonymous "boots on the ground," a phrase that is so often used to casually and carelessly refer to the thousands of soldiers performing less glamorous work.

Imagination and Prevention

Targeted killing reflects the dominance of the risk-management approach to war in several ways. First, the selectivity of the violence is meant to control the distribution of risk. The killing of a single enemy while leaving bystanders relatively unscathed shows a focusing of war on the most important enemy combatants. The goal of targeted killing is nothing less than the perfect control over how the risks associated with fighting are allocated. States employing this strategy therefore aim at total mastery of the battlefield. Precisely calibrated violence is meant to reduce the chances of unintended consequences, to narrow the possible outcomes of violence to a range in which all risks are calculable.

Second, targeted killing is anticipatory. As Beck says, when it comes to risk management, "future events that have not yet occurred become the object of current action."[23] Risks are imagined possibilities that are only faintly recognizable in the present. As I discussed previously, targeted killings are often preventive in the sense that they aim to eliminate individuals before they are able to attack, or they are punishments for past infractions. Preventive violence is only possible when managing risks; once a risk becomes a threat, the violence takes on a reciprocal character. Conversely, risk management invites preventive violence because imagining risks presupposes hyperawareness of all threatening scenarios. Rasmussen describes it as being a kind of paranoia that projects order onto the world even when none exists. "The concept of risk as the new guiding principle of strategy makes it possible to connect a number of events, policy initiatives and technological developments, which would otherwise seem random and unconnected."[24] To manage risks coming from adversaries, one must actively construct counterfactual futures that reinforce a sense of the adversary's hostility. The effort to guard against risks may therefore entrench existing hostile perceptions, foreclosing peaceful opportunities for conflict resolution and raising the likelihood that the adversary's decisions will be seen as hostile.

Third, evaluations of targeted killing are perpetually caught up in disagreements over risk calculations. One of the great controversies surrounding targeted killing is whether it is effective. When judging efficacy, the emphasis is generally on what the killings do to the target organization, though clear metrics to gauge an organization's loss of capacity are difficult to find. Should this be judged by whether the target organization disbands, whether its numbers decline, whether it is achieving its goals, or by something else entirely? Looking at the success of targeted killing in relation to terrorist organizations is especially challenging when these are usually secretive and attempt to conceal information about their own capacities. More-

23. Ulrich Beck, *World Risk Society* (Cambridge: Polity, 1999), 52.

24. Mikkel Vedby Rasmussen, *The Risk Society at War: Terror, Technology and Strategy in the Twenty-First Century* (New York: Cambridge University Press, 2006), 7.

over, it would be challenging to use a terrorist organization's performance as a measure of success when there is already debate over what it means for a terrorist organization to be successful and whether terrorism is an effective strategy.[25] With little clear evidence to draw from, the effects of targeted killing must often be judged using some outward-facing characteristic of organizations, such as their propaganda output or the number of attacks they carry out (regardless of whether these achieve anything).[26]

The debate over whether targeted killing is effective is intractable because it is undercut by the logic of risk management. It is relatively easy to determine when violence in the midst of a conflict is effective, or at least when one method is more effective than others. The same goes for pre-emptive violence. If the method of attack eliminates a visible threat, then it is effective. Because threats are concrete and specific, interactions with them can be measured. Risks, by contrast, are infinite and partly hypothetical. They exist as probabilities or as counterfactual scenarios. They cannot be decisively countered. When it works, risk mitigation is likely to go unnoticed because the imagined risk never becomes real. And when security threats become real, it is all too easy to blame mismanaged security procedures that chased imagined enemies. That is to say, the impossibility of accurately evaluating risk-management practices is built into the logic of risk. The debate over efficacy of targeted killing and other risk-management strategies cannot be settled by appeal to what the available evidence shows because the evidence is itself infected by the effects of the risk-management policies.

Fourth, targeted killing may have the boomerang effect associated with risk-management practices. By some accounts, targeted killing is counterproductive. For example, Jenna Jordan says "the marginal utility for decapitation is actually negative."[27] Stigmatization of targeted killing in the past was at least partially due to this concern. Regular armed forces thrive on order. One of the foremost reasons armed forces have historically resisted targeted killing is that the loss of key leaders could instigate a degenerate form of war in which the officers who maintain order are unable to check the excesses of the men in the ranks. As Gwynne Dyer notes, "soldiers in battle, however steady they may appear, are always a potential mob capable of panic and flight, and

25. Max Abrahms, "The Political Effectiveness of Terrorism Revisited," *Comparative Political Studies* 45, no. 3 (2012): 366–93.

26. For example, see Megan Smith and James Igoe Walsh, "Do Drone Strikes Degrade Al Qaeda? Evidence from Propaganda Output," *Terrorism and Political Violence* 25, no. 2 (2013): 311–27.

27. Jenna Jordan, "When Heads Roll: Assessing the Effectiveness of Leadership Decapitation," *Security Studies* 18, no. 4 (2009): 719–55, 723.

armies must expend an enormous amount of effort to keep them in action, beginning in basic training and continuing on the battlefield."[28] One risk, then, is that flattening the military hierarchy could replace an organized adversary obeying the norms of conventional warfare with an unorganized rabble that fights without a shared normative framework. This same concern reappears in critiques of contemporary targeted killings, as one of the fears routinely cited by those opposing this practice is that eliminating leaders could create a power vacuum in which more radical members could rise to the top. Here the risk-management practice of selectively removing specific enemies has the boomerang effect of radicalizing hostile organizations.

Names, Known and Unknown

Conventional wars are waged by anonymous masses of soldiers, the rank and file fighters whose names are rarely remembered, and who live on in the image of the "unknown soldier" who gave his life for the king/queen, the nation, or some guiding ideology. We may know the names of some great warriors who distinguish themselves in battle and become figures in literary accounts of fighting, or the generals who lead their armies into combat, but they are exceptions that prove the rule. The visibility of a handful of figures among the millions of their comrades whose names we do not know only underscores the anonymity of war. The conventional warfare of the nineteenth and twentieth centuries stands out as being especially threatening to individualism. The sheer volume of people caught up in fighting from the Napoleonic Wars, to the American Civil War, the two World Wars, and as late as the Iran-Iraq War, ensures that the vast majority of soldiers will be lost in the crowd. Lives are spent on such a scale that it is difficult to make sense of their magnitude; the killing of millions defies comprehension.

Conventional warfare is premised on anonymity. It sets combatants against each other as members of opposing groups and suppresses private interests. The soldiers do not fight because they have any personal antipathy against their enemies but because the adversaries have been deemed enemies. Once in combat, the relationship of enmity becomes more real as soldiers find themselves fighting in self-defense. The experience of violence may give rise to a more personalized sense of the conflict, especially if close friends are killed or injured. Clausewitz avers that the impersonal enmity of war can

28. Gwynne Dyer, *War: The Lethal Custom* (Carroll & Graf, 1984), 21.

affect soldiers once they have fought and feel a desire for revenge.[29] Nevertheless, this kind of enmity is capricious. There is always a chance that opposition will collapse at times when the group-based character of the fighting dissipates to reveal that opponents have no real grievances setting them in opposition. Virtually every conventional war provides instances of soldiers temporarily suspending hostilities in moments when those on both sides recognize that they do not truly hate their opponents. During the American Civil War, soldiers shared food, equipment, and medical supplies. Soldiers in the trenches of the Western Front not only paused for a truce during the Christmas of 1914 but throughout the war persisted in de-escalating hostilities and following "live and let live" norms whenever possible.[30] Michael Walzer recounts stories of soldiers who refused to attack enemies when they were vulnerable, such as when sleeping or undressed.[31] In these and many other instances, benevolence emerges because private motives push back against collective enmity.

Targeted killing breaks with the anonymity that has characterized conventional war. More than any other topic I discuss in this book, targeted killing epitomizes the individuation of war. This practice constitutes a radical shift from targeting enemies as anonymous members in a hostile collective to targeting specific people who can be identified by name or by habits revealed by prolonged observation. A successful targeted attack is usually one in which a single person is killed and bystanders go unscathed. It is only when long-range weapons become more precise and special operations forces become capable of deploying deep behind enemy lines for quick raids that this kind of precision is possible. The ultimate goal for techniques of targeted killing is to reach the point at which a single person can be killed without putting any bystanders at risk, a perfect control of risk so that its burden falls exclusively and heavily on one life.

Targeted killing is extremely controversial, but most commentators engaged in the debate seem to agree that individuation itself is positive. Proponents of targeted killing say that this strategy and the technologies used are morally superior to the mass, anonymous violence of conventional wars.[32] They defend drones and missile strikes by comparing these to blunter instruments of violence. Targeted killing then appears to be highly discriminate, or

29. Clausewitz, *On War*, 86.

30. Tony Ashworth, *Trench Warfare 1914–1918: The Live and Let Live System* (London: Macmillan, 1980).

31. Walzer, *Just and Unjust Wars*, 138–45.

32. Marcus Schulzke, *The Morality of Drone Warfare and the Politics of Regulation* (London: Palgrave, 2017).

at the very least, proportionate compared to alternative methods. Critics of targeted killing argue that attacks are not as accurate as advertised and that they inflict collateral damage.[33] By contending that attacks cannot live up to the goal of precision and that the extent of civilian harm is more extensive than is usually acknowledged, critics tacitly accept that precision is a worthwhile goal in principle and that the problems lie in execution. This signals that the logic of individuation is widely seen as normatively desirable, despite disagreements over specific methods of fighting. Those on both sides of the debate tend to agree with the view that combat ought to be restricted to the fewest participants possible. This basic commitment marks a powerful shift in norms toward an individuated conception of war.

Individuation goes beyond the moment of attack to include a host of practices associated with tracking suspects and preparing for an attack. Acquiring a target and choosing the optimal moment of attack requires extensive surveillance, during which time individuals' lives become the strategic focus. This can result in days, weeks, or even months of monitoring and careful scrutiny. The attack on bin Laden, for example, was preceded by intelligence operatives attempting to identify every person in the compound and tracking how they spent their time each day. They gave the man who turned out to be bin Laden the code name "pacer" because he habitually walked back and forth in the compound's courtyard, a habit that was suspicious because it suggested that the able-bodied man was unable to take walks outside.[34] Such a name reveals just how closely he was being watched, and it shows the effort to treat suspicious behavior as a mark of identity when the target was not known by name.

Targeted Killing and Mediation

The mediation associated with targeted killing affects the appearance of this practice for those who are involved in the operations and for the general public. For those involved, the effects of mediation are evident from how technologies of observation transform the appearance of the targets. The mediated view is epitomized by the overhead tracking shot, the omniscient view of the opponent who is not aware of being targeted. This per-

33. Maja Zehfuss, "Targeting: Precision and the Production of Ethics," *European Journal of International Relations* 17, no. 3 (2011): 543–66; Medea Benjamin, *Drone Warfare: Killing by Remote Control* (New York: OR Books, 2012).

34. Mark Bowden, *The Finish: The Killing of Osama bin Laden* (Washington, DC: Atlantic Monthly Press, 2012).

spective has been made famous by the footage of drones silently following their targets, but is not unique to them. The mediated view is endemic to new military technologies because it is one that armed forces strive to reproduce in response to the demands of maintaining total awareness of the battlefield and a freedom to kill targets at will. Satellites and manned surveillance aircraft offered a similar perspective before drones emerged. Twenty-first-century combat is therefore marked by a confluence of technologies providing this vantage point. The perspective is characterized by an asymmetry of visibility and of power, a perspective that precedes unidirectional violence that is akin to hunting. Although this is not immoral (at least by the standards of just war theory) it can feel profoundly unfair.[35]

As I showed in the opening sections of this chapter, moral objections to assassination have historically relied heavily on the assumption that such attacks will be perfidious. The attacks are described as being unfair and as degrading the shared norms that help to bring some degree of order to war. The terrorists and insurgents who figure prominently in twenty-first-century warfare are often criticized for being inherently perfidious. By refusing to mark themselves as combatants, they fight unfairly. States pursue mediated warfare as an antidote to this. Surveillance technologies make it possible to track suspected enemies long enough to ascertain whether they are hostile. Unmanned and remote weaponry make it possible to attack enemies without offering them any chance of slipping back into the civilian population. At the same time, those on the receiving end are apt to see these instruments as perfidious according to their own norms of what constitutes a fair fight. This results in competing narratives about which perspective and normative framework is right, about whether it is the nonuniformed combatants or the agents of remote death who are cheating.

One thing that is generally missing from the assessment of targeted killing's effectiveness is how attacks are received by audiences. This is methodologically challenging to measure, yet it is important to consider when attacks may be just as much meant to punish, threaten, or demonstrate capacity as to kill a particular person. By the time he was finally killed in 2011, Osama bin Laden was an increasingly marginal figure in al Qaeda. Although he was still running the organization, severe constraints were placed on his ability to coordinate attacks or to manage internal affairs. The raid that killed him may not, therefore, have caused any profound disruptions for an organization that was already used to continuing without him and that was struc-

35. Marcus Schulzke, "The Morality of Remote Warfare: Against the Asymmetry Objection to Remote Weaponry," *Political Studies* 64, no. 1 (2016): 90–105.

tured to persevere through the loss of key leaders. However, the killing certainly had a profound effect on the War on Terror, and possibly on the American presidency. The leader of al Qaeda's killing was celebrated across the United States and beyond. It was followed by a renewed wave of confidence that the United States was making some progress in the War on Terror and that President Obama could succeed where his predecessor had failed. This gave the president one of his central arguments for re-election: he was managing the conflict the right way by launching selective attacks against terrorist leaders. Thus the effects of the attack should not be gauged solely with reference to al Qaeda but also with attention to the far-reaching implications. This same point extends to other targeted killings. Targeted killing is ideally suited for constructing positive media narratives. Information about attacks and how they are carried out is closely guarded by a small number of gatekeepers. Independent fact checking is difficult. This leaves policymakers in a strong position to influence the narratives surrounding the attacks.

Rhetoric plays a vital role in mediating targeted attacks for audiences. Most victims are not as well-known as bin Laden and therefore cannot generate the same level of public jubilation at the prospect of finally making some progress against terrorism. Few members of the general public will know the names of the dozens of less central figures who have been killed in American attacks or in targeted killing campaigns launched by other countries. These people may not be well-known, yet the violence against them plays an important role in rationalizing the War on Terror. They are largely unknown, yet they are characterized as individuals with the help of status markers. When killed, they are identified with generic labels such as "al Qaeda's number two leader" or as "a senior terrorist in Yemen." Their status serves as a mark of identity to indicate the value of killing these figures to a public that is only loosely engaged in the ongoing violence. Moreover, these marks of identity are relatively fluid. An individual can only be killed once, but al Qaeda's second in command can be killed repeatedly, then celebrated each time as yet another sign of success in an intractable war. In this way, targeted killing forms a central part of the public relations campaign to sustain perceptions of success when more substantive evidence is lacking.

Killing across Borders

Imaginary borders are vital to constructing Westphalian states and their relationships. States reside in fixed territories, and statehood requires terrestrial control. National identities that are instrumental in motivating soldiers to fight are likewise closely linked to particular homelands. War between states therefore naturally follows geographical dictates, taking place within or between the territories the adversaries control. Individuals must exist in particular places, of course, but they are not tied to locations in the way states are. They are mobile and able to move beyond borders without altering their identities. When war is directed against individuals, it therefore lacks the same sense of having a natural space that exists during conventional wars that are waged against collectives. Combat can in principle follow individuals anywhere. Military and intelligence operations directed against the person must transcend boundaries if they have any hope of succeeding.

Israel's operations against former Nazis and suspected terrorists throughout the Cold War took its agents around the world and led to attacks in Lebanon, Brazil, Germany, and France, among other places. Moreover, these happened without approval from the local governments. They violated the Westphalian ideal of states respecting territorial integrity and set a precedent for conducting government-sponsored assassinations anywhere enemies could be found. The persistence of targeted killing, especially throughout the War on Terror, has entrenched this lack of respect for borders. Continued operations from Israel alongside targeted killings by the United States, Britain, and Russia take place in dozens of countries. These pose a serious threat to the future of territorial integrity as a norm governing the conduct of wars. Weak states are in a poor position to contest this imposition, and in many cases the intervening states justify their actions by appealing to disorder within the host state or claiming that state sponsorship of the target places that state outside the international community. "Rogue states" and "failed states" are treated as illegitimate and undeserving of territorial respect. Strong states may also be victims of territorial intrusion, though usually to a lesser extent. Russian attacks on Alexander Litvinenko and Sergei and Yulia Skripal in Britain are a prime example of this. In those cases, there was no overt attempt to deny the host state's claims to sovereignty, but there was nevertheless clear disrespect for sovereign authority.

Countries where enemies are found are forced into a difficult position if the intent to attack is known in advance. Permitting attacks makes the host state complicit in the violation of this norm and constitutes a partial abdication of sovereignty in the Weberian sense of having a monopoly on the legitimate use of force within the state's borders. Pakistan and Iraq fell into this position, as they permitted American and British operations against insurgents that their own militaries were unable to defeat. On the other hand, asserting territorial integrity and refusing to authorize a targeted strike amounts to defending those who are being targeted. If attacks are not known in advance, then the host state faces a different dilemma, whether to accept the fait accompli or to escalate hostilities. The response is almost always the former. Escalation after a killing or two is usually disproportionate, and the attacker may not even be a serious adversary. There was, for example, little reason for Norway to attack Israel when its agents killed an innocent waiter in Lillehammer in 1973 after he was mistaken for a terrorist. This leaves passive acceptance as the least bad alternative; it averts an escalation of violence, but at the expense of allowing attacks to go unpunished and perpetuating a style of violence that can easily transcend boundaries.

Complicating matters further is that state leaders' interests in permitting or resisting targeted killings are not uniform. These may shift according to the leader who is in power or the person being selected for attack. The United States sought permission for most of the attacks on suspected terrorists living in Pakistan under the Bush and Obama administrations, but considered it too risky to seek permission to attack bin Laden. The concern was that the Pakistani government or someone in it might alert the high-profile leader even though they had allowed the killing of less important figures. Officials in Pakistan and Yemen have allowed many of the American and British attacks to take place or even requested them, yet the same leaders generally feign noncompliance or downplay their role in the interest of preserving good public image domestically. Thus, even when support is forthcoming, the attacks are apt to appear antagonistic and thus to undermine the perceived status of state sovereignty.[36]

Of course, this all assumes that the enemy's location is known, and in many cases it is not. For every lethal operation that kills someone within a sovereign territory there are dozens of nonlethal operations to search for suspects. As Chamayou points out, "the primary task is no longer to immobilize the enemy but to identify and locate it. This implies all the labor of detection. The modern art of tracking is based on an intensive use of new technologies, combining aerial video surveillance, the interception of signals, and cartographic tracking."[37] Such total surveillance requires a constant presence within places where enemies may hide, an intrusion into physical and digital spaces that belong to others. When it is impossible to track an enemy, then war shifts further from its geographical underpinnings via efforts to monitor financial networks, business interests, and communications. Here the goal is to isolate the enemy by cutting off these outlets that act like virtual append-

36. Woods, *Sudden Justice*.
37. Grégorie Chamayou, *A Theory of the Drone* (New York: The New Press, 2015).

ages that have transnational links. Shutting down bank accounts, for example, requires concerted effort between financial institutions around the world.[38] Once again, the result is conflict beyond geographical constraints, war that is continually expanding to create a global battlespace.

Conclusion

Since the Second World War, targeted killing has been a central strategy by which states attempt to remove foreign heads of state, military commanders, and violent nonstate actors. Normatively, these attacks are best judged on a case-by-case basis. Their diversity prohibits any sweeping claims about them being either justified or unjustified under all circumstances. In general, they evince a commendable effort to comply with the principles of discrimination and proportionality, alongside a contravening challenge of undermining the status of war as a normatively distinct activity. What I have attempted to show is that they substantively change the form of war and the institutions that fight. The practice may arguably be essential in an era when so many security threats come from unconventional opponents, and it is certainly preferable to mass violence. However, narrowly focusing violence on specific enemies undermines the collective character of war and erodes the normative boundaries that distinguish war from other forms of violence. A different logic drives the attacks on enemies who are identified by name than those who are merely part of hostile groups. They are not attacked because of what group they are a part of or because of the threat they pose in at particular moment. Rather, they are moral enemies who bear personal guilt for the war itself or for some act of extreme evil taking place during the war.

The popularity of targeted killing epitomizes the twenty-first-century obsession with asserting total control over the risks associated with military operations, not only the costs to friendly personnel but even to enemies. The unprecedented effort to limit the effects of violence is morally admirable insofar as it works against the kind of indiscriminate violence that characterizes modern war, and it may even allow the rank-and-file combatants to survive conflicts. Lowering the destructiveness of combat holds the promise of making it easier to stabilize conflicts and re-establish order. Nevertheless, this individual-centric approach to war comes along with much broader

38. Juan C. Zarate, *Treasury's War: The Unleashing of a New Era of Financial Warfare* (New York: Public Affairs, 2013).

implications for the conduct of war and the structure of the military profession. As with the other processes I explore, the growing reliance on targeted killing manifests a more instrumental conception of war in which political objectives have greater primacy over the existential aspects of fighting. Precise attacks on specific enemies make it possible to centralize use-of-force decisions and to more directly ensure that attacks advance political objectives. At the same time, they undermine the Westphalian system on which our conception of politics is built.

7 • Everyone Is an Ally

Population-Centric Counterinsurgency

Since the Second World War, counterinsurgency has been the dominant mode of armed conflict. Conventional armed forces designed for pitched battles against other states were recurrently forced to fight unconventional opponents in Vietnam, Malaysia, Northern Ireland, Kenya, Afghanistan, and the Balkans. Time after time, the world's leading militaries found themselves on the losing end of these struggles or barely managing to eke out a victory. Despite its prevalence, counterinsurgency was generally treated as a deviant condition; it was a distraction from conventional war. The persistence of Cold War rivalries kept the possibility of interstate warfare alive, as did the handful of conventional conflicts that erupted throughout this era. The chance of an apocalyptic interstate war charted the direction of innovation. Shifts in strategic theory since the end of the Cold War reveal that perspectives are changing. Counterinsurgency is no longer a deviant condition; it is the usual form of war. Conventional conflicts remain possible, but the chance is more remote. Although spending on conventional weaponry remains much higher than spending on counterinsurgency technologies,[1] armed forces now devote far more resources to preparing for counterinsurgency than they did a few decades ago. This has produced substantial changes in the direction of technological, organizational, and tactical/strategic development.

The new counterinsurgency mission is radically transforming armed forces. Ethical systems and training have been redesigned. Rivalries have

1. It is important to note that the financial comparisons between conventional operations and other operations, here and elsewhere in the book, are stated in relative terms. Armed forces around the world continue to devote most of their resources to conventional operations, and will likely continue to do so for the foreseeable future. What is important is that they are also devoting more resources to preparing for other types of operations.

erupted between different conceptions of what the military profession should be. As Coker points out, "the military is becoming divided between those who still see themselves as warriors and those who see themselves as humanitarians."[2] The role of individual soldiers is changing; they are not only tasked with achieving tactical outcomes but with bringing about substantive political change. As the US counterinsurgency field manual states, "leaders must link tactical actions to strategic objectives and to the host nation's essential political goals."[3] This requires devolution of authority to lower ranks of the military hierarchy, making the decisions of individual soldiers more salient than ever. Most striking of all is that counterinsurgency is being redefined as "population-centric" warfare that prioritizes the security and interests of civilians in the battlespace. Those civilians are an objective in their own right, a "human terrain" of conflict to be won through the careful management of perceptions. This reconceptualizes the nature of political enmity into post-Westphalian terms by individuating the people within the battlespace and attempting to transform conflict into a governance activity.

The distinction between conventional war and counterinsurgency hinges on insurgents being defined as something other than state military forces. However, the difference between soldiers and insurgents can be more challenging to identify than it initially appears to be. It is possible to characterize the difference in terms of size, technological sophistication, resources, or weapons, in short in terms of material differences. Yet there are counterexamples to each of these potential defining characteristics, which makes it difficult to treat any of them as decisive. Insurgents are usually at a material disadvantage, but this is not a necessary condition. The Islamic State amassed a large stockpile of captured conventional weapons that at times gave it more destructive power than Syrian or Iraqi government forces, but it was still an insurgency. The critical distinction between states and insurgents is not material; it is normative. States belong to a special club founded on mutual recognition and assent to shared methods of fighting that constitute conventional warfare. Insurgents lack international legitimacy (they may have legitimacy among a population they claim to represent but lack recognition from established states) and refuse to fight according to the prevailing norms that states have developed to govern their own conduct.

The centrality of legitimacy as the defining difference between state mili-

2. Christopher Coker, *Waging War Without Warriors?* (Boulder: Lynne Rienner, 2002), 73.

3. *FM 3-24/MCWP 3-33/5 Insurgencies and Countering Insurgencies* (Washington, DC: Department of the Army, 2014), 1–11.

tary forces and insurgents is why counterinsurgency is often seen as a kind of degenerate warfare against enemies who do not deserve respect. From the state's perspective, insurgents lack the right to wage war and therefore lack the recognition due to legitimate adversaries. From the military perspective, insurgents look like cheaters refusing to play by the rules. From the perspective of civilian audiences watching from a distance, counterinsurgency is hardly war at all and tends to recede from the political arena as the conflicts stretch on. The wars in Afghanistan, Mali, and Gaza have become part of the background condition of life for American, French, and Israeli citizens, respectively.

Although insurgents are routinely blamed for instigating unconventional conflicts, as well as for deviating from the norms associated with conventional combat, the fault lies on both sides. War is a relational activity that takes at least two adversaries, and deviation from the established rules of this activity come from both. Belligerents can only be insurgents if they are defined as such by norms that states have developed to delegitimize nonstate actors. At the same time, states continually reinforce practices of unconventional warfare that insurgents tend to employ, as evidenced by states' reliance on the unconventional tactics used by contractors and special operations forces. There is also a natural clash between these two types of belligerents because they differ at the most fundamental levels. Insurgents are ephemeral. They lack a permanent existence apart from the insurgency. They only exist in the act of fighting. Their identity is therefore inseparable from this label. It makes little sense to talk about an insurgency taking place unless there is fighting or a threat of fighting. By contrast, states armed forces have a permanent existence and undertake counterinsurgency as one practice among others. Counterinsurgency does not define their identity but instead threatens it. For states and state militaries, counterinsurgency embodies an identity crisis and is as much a struggle to restore their own sense of normative certainty as it is a struggle against an armed opponent.

War among the People

Although the kind of guerrilla/unconventional/irregular/small war tactics employed by insurgents are as old as war itself,[4] insurgencies in the sense I am talking about are relatively new. They come into existence in relation to

4. Robert Asprey, *War in the Shadows: The Guerrilla in History* (iUniverse, 2002); Max Boot, *Invisible Armies* (New York: Liveright, 2013).

sovereign states and an international system of states in which nonstate actors are deemed illegitimate. Insurgent tactics are old, but the political act of insurgency is unique to the Westphalian system. It was only with the rise of that system that war took on a conventional appearance and that a concerted effort was made to delegitimize certain types of entities as being, by definition, unable to justifiably make war. In preceding epochs many different regime types existed concurrently, with different conceptions of sovereignty and less rigid conceptions of who was allowed to fight.[5]

Contemporary counterinsurgency practices can be traced back to nineteenth- and twentieth-century wars to establish and defend European colonial rule. These practices include heavy reliance on local auxiliaries, attempts to build friendly local governments, control of movement, encouragement of private capital investment, forced resettlement of civilians, and disaggregation of the population into antagonistic groups. Those wars were waged against nonstate entities who were outside a state system that was predominantly European, with resistance movements usually resorting to unconventional tactics based on local traditions of fighting. It is revealing that the underlying causes of contemporary counterinsurgency operations can also usually be found in this history of colonization. The various small wars throughout the post–World War II era, such as those in Algeria, Southeast Asia, Kenya, Northern Ireland, and Afghanistan had their roots in opposition to foreign rule that was established by earlier conflicts. The wars being waged now in the Middle East can be traced back to attempts to exercise control over developing countries, especially the imposition of state borders on indigenous people without adequate attention to ethnic, tribal, and religious divisions.[6]

The most important change in counterinsurgency going into the twenty-first century was the reframing of operations from being motivated by territorial expansion and control to being inspired by more high-minded political goals, such as nation building, democratization, and human security. Proposals to reapportion territory between competing factions have persistently failed in favor of less territorially focused attempts to reconcile differences without redrawing maps. The decision to build a unified Iraqi government, rather than pursuing a three-state solution, is a prime example of this. However, it should come as no surprise that so many on the receiving end of

---

5. Hendrik Spruyt, *The Sovereign State and Its Competitors: An Analysis of Systems Change* (Princeton: Princeton University Press, 1996).

6. Andrew J. Bacevich, *America's War for the Greater Middle East: A Military History* (New York: Random House, 2016).

Western assistance remain suspicious of even the most benevolent declared goals when an intervention begins. The history of counterinsurgency stands at odds with the lofty motives invoked to justify recent operations, and this historical memory is difficult to overcome in areas where those campaigns take place.

The British military has established a reputation as the world's pre-eminent counterinsurgency force, and it is widely emulated by other states. Although it has achieved mixed results, the British model appears to outperform competing doctrines developed by other states. It is particularly attractive as a guide for operations seeking to minimize risk. Because of its small size, the British military typically relies heavily on foreign security forces that simultaneously lower the cost of fighting and mitigate the image of Western imperialism. This outsources much of the risk associated with fighting, thus reducing the human and financial costs of war. The British approach prioritizes political over military goals and is guided by an overriding spirit of minimalism in all things. Minimal political aspirations that fall short of nation building. Minimal disruption of local identities. Minimal use of force. Pride in this minimalist approach became especially clear during the early stages of wars in Afghanistan and Iraq, when British soldiers critiqued their American counterparts' aggression and cultural naïveté. In a scathing analysis of the American military's performance, Nigel Aylwin-Foster writes, "COIN force must have two skills that are not required in conventional warfighting: first, it must be able to see issues and actions from the perspective of the domestic population; second, it must understand the relative value of force and how easily excessive force, even when apparently justified, can undermine popular support."[7] These maxims reveal the British aspirations and how they see their own form of counterinsurgency differing from alternatives.

The British model has become the paradigmatic example of what constitutes "population-centric" counterinsurgency. The United States' counterinsurgency doctrine went through a dramatic change between 2005 and 2007 and was heavily influenced by that ideal.[8] Endorsement of the British model came at a moment of crisis during the wars in Iraq and Afghanistan. With both countries slipping deeper into civil war and domestic opposition to the wars building, it was imperative to chart a new course that could restore hope

7. Nigel Aylwin-Foster, "Changing the Army for Counterinsurgency Operations," *Military Review* (November–December 2005).

8. Warren Chin, "Examining the Application of British Counterinsurgency Doctrine by the American Army in Iraq," *Small Wars & Insurgencies* 18, no. 1 (2007): 1–26.

of victory. Some of the most influential texts guiding this transition either focused specifically on lessons learned from British campaigns or compared these favorably against the United States' failure in Vietnam.[9] The juxtaposition of Malaya and Vietnam was especially powerful given the similarities between those conflicts and insofar as it suggested that the United States could have won in Vietnam if only it had taken a different approach. Since this transformation, the United States has more fully embraced a population-centric style of fighting that places an emphasis on providing security for civilians and building government institutions, rather than on defeating enemy forces directly. It continues to prepare for conventional war but gives greater attention than ever to training soldiers for the unique demands of counterinsurgency. As Gian Gentile observes, "the Army is so tactically oriented toward population centric counterinsurgency that it cannot think of doing anything else."[10] In fact, efforts to appeal to indigenous populations may seem to be moving beyond the British example, taking this style of counterinsurgency to an extreme. Recent British military publications even borrow extensively from American field manuals.[11]

Israel has taken a similar path in revising its conduct. The First Intifada pushed the Israel Defense Forces (IDF) beyond their core competence in wars of maneuver against the Soviet-style militaries of Arab states by obliging the IDF to contend with small-scale attacks, nonlethal attacks from rock throwers, and a hostile civilian population. At the close of the twentieth century the IDF began shifting toward a more minimalistic approach. It likewise implemented strict new ethical guidelines that were designed to constrain soldiers' use-of-force decisions and centralize control over them. Abrogation of civilians' rights continued but had to be sanctioned by higher authorities and with appeals to strategic, rather than merely tactical, interests.[12] Recent operations show a marked concern with international public opinion and fear that Israel is losing the ideological war.[13]

French counterinsurgency has likewise gone in the population-centric

9. John Nagl, *Learning to Eat Soup with a Knife: Counterinsurgency Lessons from Malaya and Vietnam* (Chicago: University of Chicago Press, 2005).

10. Gian P. Gentile, "A Strategy of Tactics: Population-centric COIN and the Army," *Parameters* 39, no. 3 (2009): 5–17.

11. For example, see British Army Field Manual, vol. 1, part 10, *Countering Insurgency* (October 2009).

12. Marcus Schulzke, *Pursuing Moral Warfare: Ethical Theory and Practice in Counterinsurgency Operations* (Washington, DC: Georgetown University Press, 2019).

13. Ron Schleifer, "Jewish and Contemporary Origins of Israeli *Hasbara*," *Jewish Political Studies Review* 15, no. 1–2 (2003): 123–53.

direction. Failures in Southeast Asia and Algeria taught a harsh lesson about the impossibility of winning unconventional wars using conventional methods. Writing in 1961, Roger Trinquier, who is among the most influential figures in French counterinsurgency theory, says "we know that the *sine qua non* of victory in modern warfare is the unconditional support of a population."[14] He adopts a population-centric perspective, though one that is far from gentle. It harkens back to colonial operations, advocating forced relocation, re-education, local defense forces, and even torture. Strategic thinking in the 1980s and 1990s was caught up in a search for alternative tactics and showed a newfound concern for protecting indigenous populations rather than simply dominating them. It preserved Trinquier's insight that the population matters, while searching for more attractive methods. French operations in Mali opened with selective strikes involving special operations forces, air assets, and a limited number of regular ground personnel. It then settled into a protracted security mission that privileged political development and the empowerment of local security forces.[15]

The Russian military attempted to retake the breakaway states of the Caucuses with a conventional military invasion during the 1990s, resorting to extremely harsh methods like shelling urban areas and executing family members of suspected insurgents. These measures resulted in humiliating defeats and prompted a search for different approaches.[16] Russia persisted in launching devastating conventional attacks during the Second Chechen War, but it was ultimately able to secure the contested areas by adopting some of the methods characteristic of population-centric warfare. In particular, it benefited from using local proxies like Akhmad and Ramzan Kadyrov to maintain order. Subsequent operations in Syria and Ukraine suggest that it has continued investing in this strategy. In the latter case, it went further by relocating civilian populations to create more solidly pro-Russian constituencies.[17] Russia still falls short of North American and western European countries when it comes to civilian outreach, but the rapid adjustment to strategies of reduced violence and population control mark an important shift.

---

14. Roger Trinquier, *Modern Warfare: A French View of Counterinsurgency* (Fort Leavenworth, KS: US Army Command and General Staff College, 1985), 5.

15. Michael Shurkin, *France's War in Mali: Lessons for an Expeditionary Army* (Santa Monica, CA: RAND, 2014).

16. John Russell, *Chechnya—Russia's 'War on Terror'* (New York: Routledge, 2007).

17. Ilan Berman, "How Russian Rule Has Changed Crimea," *Foreign Affairs*, July 13, 2017, https://www.foreignaffairs.com/articles/eastern-europe-caucasus/2017-07-13/how-russian-rule-has-changed-crimea

The Emergence of New Norms

Historically, many norms governing the use of force during combat arose from a sense of reciprocity between belligerents who had a shared interest in limiting certain aspects of violence.[18] Restrictions on what weapons could be used, when battles could be waged, and who should be taken prisoner were mutually beneficial. Such conventions tended to be strongest among elite warriors, who shared kinship links and used norms to collectively guard their status from those lower in the socioeconomic hierarchy. Chivalry is a prime example of this. That shared code of values linked the warrior elites of Europe throughout the Middle Ages, giving opposing belligerents a common reference point about how they should fight.[19] By promoting the capture and ransoming of noble opponents, rather than killing them, it helped to minimize casualties among elites and to ensure that they would generally not suffer such high losses as to open a political vacuum that could be filled by others. Restrictions on weapons like the crossbow and a general sense that ranged weapons were cowardly and unsuitable for real warriors helped to ensure that fighting remained a specialized skill dominated by those who made hand-to-hand combat their profession. Moreover, norms focused primarily on the treatment of enemies. Before the twentieth century, norms protecting civilians were weak and routinely discarded. Elite codes of conduct, as well as just war theory, encouraged mercy for civilians, but failed to identify them as a protected class of people who could not be attacked. Only certain segments of the civilian population, such as priests, were widely considered to have absolute immunity.[20]

Throughout the nineteenth and twentieth centuries, belligerents made unprecedented efforts to strengthen and formalize the norms of war. The Hague Conventions of 1899 and 1907 and the Geneva Conventions of 1929 stand out as some of the most important agreements in the development of international law, yet these continued the trend of focusing on the protection of soldiers. The agreements talked extensively about what weapons and tac-

18. Norms also came from religious doctrine, but I set these aside here because these norms tended to only pertain to wars between belligerents of the same religion and were not general rules of conduct. Moreover, they likewise showed little concern for civilians until the twentieth century.

19. Maurice Keen, *Chivalry* (New Haven: Yale University Press, 2005).

20. James Turner Johnson, *Just War Tradition and the Restraint of War: A Moral and Historical Inquiry* (Princeton: Princeton University Press, 1981); Helen M. Kinsella, *The Image Before the Weapon: A Critical History of the Distinction Between Combatant and Civilian* (Ithaca: Cornell University Press, 2011).

tics could be justifiably used but paid relatively little attention to the treatment of civilians. This was then reflected in the subsequent conduct of the total wars between conventional belligerents, in which civilians were killed in shocking numbers. The Second World War marked a turning point, as both sides targeted civilians and used terrorist tactics premised on coercing opponents through the infliction of intolerable losses. This was the terrible legacy of the growing popular involvement in "people's wars" since the Napoleonic era. With entire populations being mobilized, it was logical for belligerents to extend the scope of fighting to encompass entire populations: to fight external wars of extermination and to pursue extermination or mass internment as a domestic security agenda.

Insurgents are frequently blamed for endangering civilians, and they do indeed act in many ways that threaten civilians' protected status. In particular, they have a penchant for blending in with civilians or using them as shields. And yet it is also important to recognize that we owe much to insurgencies for raising concerns of civilian protection. The other response to people's wars, which developed in opposition to the total war practice of extermination, was to co-opt civilians and transform them into a base of support. Throughout the early twentieth century, insurgencies reached into civilian populations in this way. The century's most influential theorists of revolution shared a desire to create mass popular forces. Following the Second World War, the waves of nationalist and communist uprisings took popular mobilization to unprecedented levels, fueled by pragmatic sensibilities about what was necessary for victory and by ideological influences that promoted national unity and the socioeconomic elevation of workers. Mao Tse-tung codified the insurgent faith in mobilizing popular support and using civilians as a source of power. From that point on, control of civilians would be a key objective for virtually every insurgency, and state policies of civilian victimization would be counterproductive in almost every case. Regular militaries gradually learned this lesson, often adopting Mao's recommendations for their own purposes.[21]

Civilians were envisioned as potential targets of interstate fighting during the Cold War, as evidenced by the nuclear doctrine of mutually assured destruction and the design of conventional munitions so powerful that they would necessarily inflict civilian casualties in the heavily populated European theater, where a confrontation seemed most likely. Throughout the Cold War, states fighting insurgencies continued to target civilians, yet they

---

21. Mao Tse-tung and Samuel B. Griffith, *On Guerrilla Warfare* (North Chelmsford, MA: Courier Corporation, 2005).

began sporadically appealing for local support. For example, the United States conducted mass bombings throughout Vietnam, but it also attempted to build a functioning army in South Vietnam, cooperated with Montagnard tribes, and experimented with embedding soldiers in local communities. There were tentative steps toward a more population-centric approach.[22] Moreover, this was the era of British victories in Malaya and Kenya, which would become the primary cases used to demonstrate the efficacy of population-centric operations. The practice of population-centric counterinsurgency therefore began to take shape under the combined influence of failing conventional approaches, the influence of insurgents' own writings, and the limited successes of population-centric approaches in certain areas.

Growing recognition that population-centric strategies were more effective for countering insurgencies coincided with the strengthening of norms related to civilian protection. Revisions to international law following the Second World War gave greater attention to civilians and to nonuniformed combatants. In particular, the Fourth Geneva Conventions of 1949 established clearer rules protecting civilians.[23] The Cold War was therefore a time of practical necessities and moral/legal imperatives reinforcing each other. Armed forces' success or failure depended heavily on whether they recognized these changing circumstances and were able to adapt. Unfortunately, the momentum of traditionalism and the continued threat of conventional war hindered adaptation. It was not until the chance of interstate warfare diminished that the world's most powerful militaries began to fully embrace a new mindset of restricting their lethal operations and showing greater concern for civilian welfare. Although this is manifest in a range of different counterinsurgency strategies, it is perhaps best summed up by General Stanley McChrystal's theory of "courageous restraint," which calls for soldiers to minimize force, even to the extent of undergoing greater personal danger, to secure the better relations with civilians that will help to achieve political victory.

Mastering the Human Terrain

Ground fighting in conventional wars has historically been geographically oriented. Opposing forces struggle to control a given plot of land and iden-

---

22. Lewis Sorley, *A Better War: The Unexamined Victories and Final Tragedy of America's Last Years in Vietnam* (New York: Harcourt Brace & Company, 1999).

23. Convention (IV) Relative to the Protection of Civilian Persons in Time of War, Geneva, August 12, 1949.

tify each other in terms of which territories are under their respective control. The enemy can be located in a particular area and brought to battle by invading that place. This ordering of war through oppositional spaces is particularly clear when conventional wars create front lines that shift back and forth as the conflict unfolds. Analyses of these wars typically focus on how the lines moved and how the victors were ultimately able to take the enemy's ground. Yet even in the Second World War, which in many ways epitomizes conventional warfare, there were hints of the changes to come. Although they did not play a decisive role, partisans throughout Europe and Asia made major contributions by waging unconventional wars against the Axis forces and tying hundreds of thousands of soldiers down in garrison and policing roles. The lesson of unconventional fighting in Yugoslavia, Poland, France, and Southeast Asia was clear: having a preponderance of force in a territory does not guarantee control. When the population is not content with its governance, it may continue to resist.

Conventional forces routinely failed in unconventional conflicts throughout the Cold War because they continued to devote most of their resources to dominating territory and stopping a geographical spread of influence. During the decades of rivalry, no propaganda film was complete without showing a map of the creeping line of Soviet expansion or the coalition of Western states arrayed against it. At times, conventional forces did confront each other in struggles for territorial control, as they did during the Korean War and the Iran-Iraq War, and in these instances victory was pursued through the effort to take and hold territory. Nevertheless, the same militaries that performed well in conventional operations struggled, and usually failed, to rethink their objectives and to adjust their operations accordingly. They sought to capture territory, often rushing in to seize control of it as quickly as possible, only to find that territorial mastery meant little to a determined enemy. Even as late as the post-9/11 campaigns in Afghanistan and Iraq, conventional forces committed the hubris of equating geographical dominance with political victory.

Much of the strategic innovation during the twentieth century came from the British military, as it was in British colonial operations that the population-centric approach took shape as an alternative, or supplement, to the use of conventional military power. In some instances, capture of the terrain was enacted literally. Counterinsurgency forces tended to be most successful when they physically separated civilians from insurgents by displacing the former into internment camps and creating secure zones that would deny insurgents access to their civilian supporters. Early versions of this approach can be seen from the Boer War, when the British pursued two forms of population governance: constructing a long chain of mutually supporting guard posts to restrict movement, and transporting Boer civilians into camps. Similarly, success during the Malayan Emergency and the Mau Mau uprising in Kenya came when the British and local government forces established strict control over the local civilian population to cut insurgents' logistical support.[24] It is in these operations that we can see a shift in thinking about civilians, from civilians-as-targets to civilians-as-capturable-terrain. These examples also show that the population-centric approach to counterinsurgency is not necessarily easier on civilians or more ethical. Historical attempts to assert control over the human terrain have led to major atrocities, with tens of thousands of civilians dying from disease or malnutrition when they are moved or interned. In these cases, the overt civilian victimization of deliberate attacks is replaced by the indirect exposure to dangerous conditions.

The treatment of civilians as terrain to be controlled is a defining characteristic of the population-centric approach, both in theory and in practice. During the wars in Iraq and Afghanistan, the US and UK militaries began to classify civilians as the "human terrain." The 2006 counterinsurgency field manual states that "The cornerstone of any COIN effort is establishing security for the civilian populace."[25] Counterinsurgency practices are often described with abstract spatial metaphors that indicate the deracination of these campaigns from physical territory. For example, Emile Simpson, a former British officer, says that "The control of political space is as important, if not more important, than controlling physical space."[26] It is revealing that space is still being treated as something to control or to win. Space is still the objective, it is just not solely physical. In the case of political space, the terrain expands beyond physical embodiment in the land or in people, becoming a struggle over narratives that imbue war with meaning. Controlling the human terrain is largely a perceptual matter. Civilians are treated as an audience that can be won through persuasion. They are potential allies, or they are at least weakly committed to the insurgents' cause and open to defection. There is a preoccupation with acting in ways that will create the best possible view of the war for those audiences, in the hope that the victory over insurgents will come from changes in civilian attitudes.

The United States has attempted to remake the landscape around people

---

24. Nagl, *Learning to Eat Soup with a Knife*; Huw Bennett, *Fighting the Mau Mau: The British Army and Counter-Insurgency in the Kenya Emergency* (Cambridge: Cambridge University Press, 2013).

25. *FM 3-24/MCWP 3-33.5 Counterinsurgency* (Washington, DC: Headquarters Department of the Army, 2006), 1–131.

26. Emile Simpson, *War from the Ground Up* (Oxford: Oxford University Press, 2012), 6.

to manage interactions. This is in a sense a reversal of the old interest in geography as the terrain of battle, which makes geography one tool for organizing the human terrain. US forces constructed a three-mile-long wall through Baghdad to limit interactions between Shiite and Sunni communities. They established a network of observation posts along the roads to monitor movement and attack insurgents placing IEDs. Traffic control points were established throughout the country to regulate the movement of goods and people. In a few cases the US military pursued policies of temporary population resettlement, as when it evacuated Fallujah to create a setting for a conventional assault on the city.[27] Israel has employed the same practices, relocating Palestinians into specific areas of the country and limiting their movement with walls and checkpoints. They likewise relinquish control of particularly contentious areas like Gaza and the West Bank, recognizing that governance of the people living there is a vital national security concern but that possession of the land is not. In these ways, security forces remake the terrain to fit the demands of population control, privileging the human terrain over physical terrain.

Everything Is Political

Population-centric counterinsurgency is a style of fighting that emphasizes the primacy of political goals. A core guiding principle is that victory cannot come from battlefield successes alone. Insurgents may be defeated in every engagement and still manage to win the overall war because of their superior political acumen, and especially their support from the indigenous population. In the past, it may have been possible for Western forces to so thoroughly dominate their adversaries that resistance became unthinkable, or to simply intimidate insurgents through mass killings. However, such conduct is increasingly rare. Atrocities committed by state militaries are met with harsh responses from the international community and are apt to provoke resistance from local people. The decline of aggressive counterinsurgency is evident from how Russia used brute force during the wars in Chechnya, then shifted to less direct tactics elsewhere in the Caucuses and in Ukraine. Twenty-first-century counterinsurgency warfare marks an important turning point in the over-

27. John R. Ballard, *Fighting for Fallujah: A New Dawn for Iraq* (Westport, CT: Praeger, 2006); Patrick O'Donnell, *We Were One: Shoulder to Shoulder with the Marines Who Took Fallujah* (New York: Da Capo Press, 2007).

all orientation of war. It signals the primacy of the political over the military. Combat is made less important than diplomacy and reconstruction, even to the extent that it must be avoided in the interest of maintaining good relations with local civilians. The military is itself transformed into a political organ that must be more concerned with policing, training, reconstruction, and even community outreach programs than with fighting. Military goals are not distinct from political goals. In effect, war is not only an extension of politics with a life of its own, as Clausewitz says, but an extension that never distinguishes itself from politics.

The ascendancy of political over military imperatives entails a transformation of the military profession and a reordering of soldiers' identities. A defining feature of the military profession is that individual soldiers and the military as a whole are apolitical, or at least not explicitly partisan. Soldiers are supposed to be disinterested servants of the state who obey the policy objectives they are given. Counterinsurgency politicizes militaries and their soldiers in two ways. First, it forces individual soldiers to go beyond making decisions about the use of force to making political decisions within the affected country. They do not merely fight but also participate in building infrastructure, governing institutions, and even national identities. Soldiers become diplomats, politicians, and civil engineers. Second, counterinsurgency places armed forces in the awkward position of having to advance partisan causes that will lead them to run afoul of the opposition perspective. Throughout the US war in Iraq, antiwar voices clamored for a speedy withdrawal. For many, the only salve for a misguided war was to end it quickly and leave Iraq to govern itself. The more moderate antiwar position was to recognize the necessity of continuing the fight, but with a dramatically reformulated goal like dividing Iraq into three states. The military's efforts to win the war and to keep Iraq intact in the process therefore became partisan decisions. When General Petraeus went before Congress to testify about the necessity of the 2007 troop surge, the liberal organization MoveOn responded by running a full-page ad in the *New York Times* with the title "General Petraeus or General Betray Us?"[28] Prominent figures on the left generally decried what they saw as an escalation of force to support failing Bush administration policies.

Politicization within the conflict area is unavoidable given the deep involvement between soldiers and local governing institutions. Soldiers

28. Kevin Anderson, "General Betray Us v Phony Soldiers," *The Guardian*, October 5, 2007, https://www.theguardian.com/world/deadlineusa/2007/oct/05/generalbetrayusvsphonysol

sometimes fight pitched battles against insurgents, yet they more commonly conduct security patrols through populated areas, operate checkpoints designed to intercept insurgent traffic, and search homes. They are also charged with performing civil-military affairs work, which comes in such diverse forms as giving candy to children and assisting in the reconstruction of power lines. As Gilmore observes, "concern for the well-being of local populations, a step away from a strictly military security focus and the presence of a distinctive development component in US counterinsurgency doctrine imply a transition toward something more akin to human security."[29] The British counterinsurgency manual identifies a triad of counterinsurgency tasks: development, governance, and security.[30] It even describes soldiers as having hybrid roles as "soldier-diplomat," "soldier-administrator," and "soldier-politician."[31]

The shift between combat and administrative roles is apt to produce cognitive dissonance. Soldiers often object to acting like police officers and being held to similar standards even when they may have to reprise the role of soldier at a moment's notice.[32] From the soldier's perspective, the normative degeneracy of counterinsurgency infects their own worth and causes them to suffer a drop in status. Many soldiers, especially those in combat arms units, seek war for a chance to become warriors, to be heroes in the eyes of their fellow citizens and to live on in the pages of history. Instead what they find is a largely unreceptive population that is ignorant of their campaigns or opposed to them. They face enemies who are inherently unequal by virtue of their lower status. Much as Homeric heroes and medieval knights might suffer a loss of prestige for fighting against lesser opponents rather than taking on their social equals, soldiers in state military forces fail to earn the prestige of their forbears and even take on a tinge of the insurgent's stigma. As French notes, "It is easier to remain a warrior when fighting other warriors."[33] The irony is that much of this stigmatization of insurgents comes from military professionals. They define themselves as being distinct from others, and potentially better than others, by virtue of their special military identity. The

---

29. Jonathan Gilmore, "A Kinder, Gentler Counter-Terrorism: Counterinsurgency, Human Security and the War on Terror," *Security Dialogue* 42, no. 21 (2011): 21–37, 32.

30. British Army Field Manual, vol. 1, part 10, *Countering Insurgency* (October 2009), 1–4.

31. British Army Field Manual, vol. 1, part 10, *Countering Insurgency* (October 2009), 3–5.

32. Schulzke, *Pursuing Moral Warfare*.

33. Shannon E. French, *The Code of the Warrior: Exploring Warrior Values Past and Present* (New York: Rowman & Littlefield, 2003), 241.

warrior identity can only be conferred by mutual recognition by other warriors, which is why so many soldiers acknowledge a respect, or even admiration, for their enemies. So long as counterinsurgency remains the predominant form of war and insurgents are defined in opposition to regular armed forces, soldiers cannot gain this recognition. This places their own identities in jeopardy.

Fighting against unconventional opponents and taking on roles that are beyond the scope of how the military profession has traditionally been defined undermines confidence in existing conceptions of what it means to be a soldier. It weakens the sense of a distinct military profession with unique roles. Bell goes so far as to characterize counterinsurgency as a "civilianisation of warfare."[34] This is partially true. Counterinsurgency presents a dilemma. The majority of soldiers are civilianized as they are channeled into work associated with reconstruction and security. As I pointed out in earlier chapters, soldiers in special operations roles continue to be highly militarized, albeit in a distinctive new way because they continue combat in an unconventional manner. This means that the erosion of the military profession does not take a single trajectory. Instead, it moves in multiple contrary directions, which are already manifest in the division of conventional and special forces.

Entire units may find that their specializations are no longer relevant in counterinsurgency operations and that they have to be redeployed in roles for which they were not trained. Those in field artillery and air defense artillery often suffer in this respect. Without large concentrations of enemy fighters or any aircraft to target, they can only participate in the fighting when they are forced into new types of work. American artillery units deploying to Iraq and Afghanistan halted training programs for high-intensity combat and focused on more limited roles like suppressing snipers. One report from an artillery unit that served in Iraq in 2007 describes how the soldiers were tasked with guarding detainees alongside Air Force and Navy personnel.[35] Even the long-term future of artillery is uncertain. "Artillery . . . often referred to as the 'King of Battle' is now thought of as a 'dead branch walking' as its role in the COIN/IW is greatly diminished relative to conventional warfare."[36] Some proposals call for the merging of associated units that pri-

34. Colleen Bell, "Civilianising Warfare: Ways of War and Peace in Modern Counterinsurgency," *Journal of International Relations and Development* 14, no. 3 (2011): 309–32, 312.

35. Christopher R. Vegas, "Fire and Maneuver: Versatility in COIN," *Fires* (October–December 2008): 34–37.

36. Robert M. McNab and Richard L. Scott, "Nonlethal Weapons and the Long Tail of Warfare," *Small Wars & Insurgencies* 20, no. 1 (2009): 141–59, 145.

marily have conventional war roles, such as by combining field artillery with air defense artillery. These units certainly still matter, but their efficacy is reduced and they are at greater pains to remain relevant.

Another revival from the colonial history of small wars is the practice of enlisting assistance from academia. US counterinsurgency strategy, especially after the shift to a more population-centric approach in 2005, has been assisted by academic research.[37] Israel's ethics code was primarily written by Asa Kasher, a professor at Tel Aviv University. Academics have even worked on the front lines, most notably through the US military's creation of Human Terrain Teams made up of civilian researchers. These teams, which were active from 2007 to 2014, were designed to map the human terrain using social science research tools that were beyond the scope of military training. The result was a collective civilian and military endeavor to counter insurgents. Efforts to learn about local populations and to show respect for cultures elevate the role of academic research, especially anthropology.[38] This acknowledges the limits of military expertise when it comes to civil engagement and introduces yet another route of civil-military cooperation that opens gaps in the boundaries of the profession.

Warriors and Social Workers

The goal of conventional battle is usually to bring overwhelming strength against an enemy to establish local superiority. Mass and unity of effort are keys to winning. This is why pitched battles until the twentieth century were typically fought by concentrated formations of soldiers. This only began to change when more powerful artillery, rifles with longer ranges and higher rates of fire, and machine guns made dense formations untenable. War in the late nineteenth and early twentieth centuries was characterized by rapid increases in "complexity, mobility, and dispersion."[39] These processes made it impossible to impose authoritarian top-down control and forced a devolution in decision making to lower ranks. Greater mobility made it essential for commanders in the field to have the freedom to improvise. Complexity made micromanagement slow and impractical. Dispersed

37. Martin G. Clemis, "Crafting Non-kinetic Warfare: The Academic-Military Nexus in US Counterinsurgency Doctrine," *Small Wars & Insurgencies* 20, no. 1 (2009): 160–84.

38. Montgomery McFate, "Anthropology and Counterinsurgency: The Strange Story of Their Curious Relationship," *Military Review* 85, no. 2 (2005): 24–38.

39. Martin Van Creveld, *Technology and War: From 2000 B.C. to the Present* (New York: Macmillan, 1991).

soldiers had to be more self-reliant, especially when lacking radio communication. Taken together, these processes made lower-ranking soldiers more important as decision makers.

These processes have continued into the twenty-first century, to the point that a smaller number of soldiers stretched across massive battlespaces must be able to make quick, independent decisions, sometimes with minimal guidance beyond a commander's overall mission intent. And with political goals being prioritized in counterinsurgency, soldiers' decisions take on greater weight than ever. Their decisions are not solely military in scope; when the military mission prioritizes political outcomes, the decentralization of authority and empowerment of individuals takes on a political character. Soldiers are not merely soldiers. They are also diplomats, social workers, police officers, and community organizers. They are "strategic corporals" whose actions on the tactical level have strategic and political impact.[40] As Simpson points out, "relatively junior commanders find themselves making decisions which although nowhere near as significant in scale as 'strategic' decisions made by those who have strategic authority, nonetheless have a directly political quality."[41] He also explains that "political considerations now drive operations even at the lowest level of command: the military dimension of war is pierced by political considerations at the tactical level."[42]

This way of thinking is evident in population-centric doctrine. The 2006 US counterinsurgency field manual states that "Tactical actions thus must be linked not only to strategic and operational military objectives but also to the host nation's essential political goals."[43] Tactical, operational, strategic, and political matters blur together to such an extent that these cease to exist as separable dimensions of conflict and instead become different aspects of a continuum of decision making. The devolution of authority and profound political importance of soldiers' actions is reflected in military ethics training programs, which have become far more elaborate and central to military education in the twenty-first century.[44] It has likewise given rise to more cul-

40. Charles C. Krulak, "The Strategic Corporal: Leadership in the Three Block War," *Marines Magazine* 28 (1999), http://www.au.af.mil/au/awc/awcgate/usmc/strategic_corporal.htm

41. Simpson, *War from the Ground Up*, 93.

42. Simpson, *War from the Ground Up*, 6.

43. *FM 3-24/MCWP 3-33.5 Counterinsurgency* (Washington, DC: Headquarters Department of the Army, 2006), 1–156.

44. Paul Robinson, Nigel De Lee, and Don Carrick, eds., *Ethics Education in the Military* (Hampshire: Ashgate, 2008); James Connelly, Don Carrick, and Paul Robinson, eds., *Ethics Education for Irregular Warfare* (Burlington, VT: Ashgate, 2009).

tural awareness education aimed at smoothing interactions between soldiers and indigenous civilians. Soldiers must be trained in an array of skills that will allow them to make the right decisions on their own initiative. The British counterinsurgency manual says that "the soldier who has to fight the insurgent is the soldier who will have to get involved with local politics, local economic development projects, and programmes to improve local services and create and sustain conditions for national development."[45] Here it is clear that the devolution of authority goes across all aspects of the counterinsurgency mission, entrusting individual soldiers with unprecedented autonomy.

The devolution of authority can be liberating; it can also cause an alienating sense of isolation from a larger collective. Reflecting on his time in Iraq, Col. Nathan Sassaman says that "I discovered too late that we were not really a nation at war in Iraq; too often, in fact, we weren't even an army at war. We were a handful of battalions at war, led by an even smaller group of officers."[46] He reflects on the overriding sense of incoherence he experienced because of this fragmentation. Frank Ledwidge offers similar opinions from the British perspective. Both express the frustrations of being given a daunting mission without adequate resources and left to fend for themselves. Even commentators who have a more optimistic perspective on the enormous political responsibility note that devolved authority can produce a wide variety of counterinsurgency practices between different areas of responsibility, to such an extent that progress toward mission objectives varies considerably depending on the region.[47] Thus, population-centric counterinsurgency conducted by armed forces that are spread thinly has caused an individuation within the ranks.

The disaggregation of groups into small unit and individual contributions also takes place with respect to the insurgents. Within strategic narratives they are described less as members of a coherent collective than as a loose agglomeration of people with competing goals and interests. David Kilcullen, one of the leading intellectual influences on the US counterinsurgency doctrine, says that most insurgents are "accidental guerrillas" who do

---

45. British Army Field Manual, vol. 1, part 10, *Countering Insurgency* (October 2009), 3–22.

46. Nathan Sassaman and Joe Layden, *Warrior King: The Triumph and Betrayal of an American Commander in Iraq* (New York: St. Martin's Griffin, 2008), 3.

47. Thomas E. Ricks, *The Gamble: General David Petraeus and the American Military Adventure in Iraq 2006–2008* (New York: Penguin Press, 2009); Peter R. Mansoor, *Surge: My Journey with General David Petraeus and the Remaking of the Iraq War* (New Haven: Yale University Press, 2013).

not have a strong stake in the main political contest.[48] Their fight is for local concerns, such as tribal rivalries and vengeance. This means they can be co-opted by thoughtful commanders who are able to offer assistance. Ledwidge supports this, saying that "Most people in the Afghan conflict are really actors in their own right, and act according to their own interests, as opposed to that of a given side."[49] By this logic, victory does not come from defeating an enemy collective outright, as in conventional operations. The goal is instead to disaggregate the enemy until the core fighters are isolated and vulnerable. Academic research on insurgencies supports this conclusion. For example, Stathis Kalyvas argues that most fighters are motivated by local or private concerns, rather than the main political objectives, and that "individual and local actors take advantage of the war to settle local or private conflicts often bearing little or no relation to the causes of the war or the goals of the belligerents."[50]

This narrative of insurgents as heterogeneous enemies is essentially the individuation of insurgency to the point that insurgencies lose coherence. It prioritizes a multifaceted attempt to fight committed insurgents while also appealing to the interests of those who are weakly aligned. The latter effort is largely waged by attempting to influence individuals' perceptions. The weakly aligned insurgents should be persuaded rather than killed. "A group of 'accidental guerrillas' may support the Taliban, but one should not be seeking their defeat; rather one should be seeking to pull them out from the sway of the Taliban."[51] Ideally, the accidental guerrillas may even become allies in the struggle, thereby carrying out Sun Tzu's advice of drawing strength from the enemy's resources. This demands attention to individual grievances and efforts to resolve those grievances on a case-by-case basis. Thus, for the insurgents as for the soldiers, the war effort shifts from being about collectives to a series of individual decisions that political outcomes emerge from.

---

48. David Kilcullen, *The Accidental Guerrilla: Fighting Small Wars in the Midst of a Big One* (New York: Oxford University Press, 2009).

49. Frank Ledwidge, *Losing Small Wars: British Military Failure in Iraq and Afghanistan* (New Haven: Yale University Press, 2011), 3.

50. Stathis N. Kalyvas, "The Ontology of 'Political Violence': Action and Identity in Civil Wars," *Perspectives on Politics* 1, no. 3 (2003): 475–94, 476.

51. Simpson, *War from the Ground Up*, 100.

Paradoxical Risks

The individuation of insurgents is part of a risk-management strategy and is often explicitly couched in that language. As the British counterinsurgency manual says, "Separation of the insurgent from the population psychologically is perhaps the most complex aspect of separating the insurgent from his support but it must draw due attention as it offers a more enduring result with less risk of unintended consequences if achieved."[52] Such language acknowledges that there is no sure victory when dealing with insurgents. The best that can be hoped for is a strategy that allocates risk in the optimal way. It is revealing that the risk identified here is that of unintended consequences. Preoccupation with counterproductive actions illustrates reflexive thinking that is characteristic of risk aversion.[53] It reveals the anxiety in the realization that any action could have perverse outcomes. This thinking is underscored by how the US and UK counterinsurgency doctrines emphasize paradoxical outcomes with guidelines like "the more you protect your force, the less secure you are," "the more force you use, the less effective you may be," "sometimes doing nothing is the best reaction."[54] Such maxims read like a Zen koans. The risk calculations are intractable and resist any conclusive resolution. Guidelines are contingent. They are not fixed truths. In the end, soldiers are left to adapt as best they can to the changing field of risks with the reflexive attention to all future possibilities as the primary guide.

The struggle to win the human terrain shifts priorities away from conventional military objectives. In conventional warfare, risk calculations primarily revolve around force protection and mobilizing firepower against the enemy. The population-centric goal of capturing the human terrain expands priorities beyond force protection to human security more broadly. Efforts must be made to limit insurgent attacks on civilians and to prevent violence between competing insurgent factions that may exacerbate domestic insecurity. The multiplicity of actors involved therefore compounds the risk calculations beyond what they would be when war is conceived as a contest between two opposing collectives seeking to destroy each other.

---

52. British Army Field Manual, vol. 1, part 10, *Countering Insurgency* (October 2009), 3–12.

53. Anthony Giddens, *Modernity and Self-Identity: Self and Society in the Late Modern Age* (Cambridge: Polity Press, 1991).

54. British Army Field Manual, vol. 1, part 10, *Countering Insurgency* (October 2009), 3–22.

Expanding the scope of risk calculations in this way introduces tradeoffs between security for soldiers and civilians. Any interaction between the two must be governed by use-of-force rules that shift the burden of risk onto one group or the other. Permissive rules of engagement allow soldiers to protect themselves more effectively but at the cost of increasing the risk of accidental attacks on civilians. Conversely, prioritizing civilians' lives may raise the number of military casualties and foster dissent among soldiers who feel that they are being inadequately protected. This dilemma is often explicitly acknowledged in counterinsurgency doctrine. For example, the 2014 US counterinsurgency manual says that "Soldiers and Marines must accept some risk to minimize harm to noncombatants. Accepting prudent risk is an essential part of the warrior ethos and an obligation of honorable service."[55] The language of risk here is telling. Soldiers are asked to enter into the risk calculations themselves, voluntarily taking on a greater share of immediate risks for the promise of long-term payoffs when it comes to building local support for the mission.

How exactly soldiers are supposed to make these risk calculations is never fully spelled out. Gilmore rightly says of the US counterinsurgency manual that "Although *FM 3-24* emphasizes the necessity of a more restrained use of force, it is less clear about how soldiers trained and equipped primarily for high-impact war-fighting will, in practice, switch fluidly from one posture to another."[56] When war is conceived as an exercise in risk management, this is the only possible outcome. No instructions can be conclusive. They can only expound on the paradoxical logic governing the situation and provide tentative assistance, while leaving the ultimate decision up to the individual actors involved.

There is also a broader sense in which population-centric counterinsurgency introduces new risks. The more attention is given to human security and training soldiers in law enforcement activities, the more soldiers are drawn away from conventional war-fighting missions. It is unrealistic to try to create a military that is equally effective in all conflict settings. Decisions have to be made about what risks to prioritize and how to plan for the future. Investment in counterinsurgency seems like the best strategy at the moment, but interstate conflict remains a possibility and becomes increasingly risky as more resources are drawn away from preparation for it. With each invest-

55. *FM 3-24/MCWP 3-33/5 Insurgencies and Countering Insurgencies* (Washington DC: Department of the Army, 2014), 1–11.

56. Gilmore, "A Kinder, Gentler Counter-Terrorism," 27.

ment in population-centric approaches comes doubts about whether armed forces will be able to take on conventional war-fighting missions in the future.[57] This dilemma is one of the reasons why armed forces are pursuing modular capabilities: weapons that can be adapted to multiple roles, vehicles that can be transformed, and units that can be assembled into different configurations. Modularity is adaptability. It is a hedge against risk. A modular force has greater potential to quickly transform to meet whatever risk scenario becomes real.

Knowing the Audience

The image of a gentle counterinsurgency is largely illusory; violence is unavoidable. This is particularly true when states attempt to assert control over civilian populations by moving them into camps where they are susceptible to disease and malnutrition.[58] Sometimes the violence is inflicted intentionally; sometimes it is accidental. For example, the British decision to withdraw from southern Iraq in 2007 set the conditions for brutal repressions by the Mehdi Army, which ultimately required American and Iraqi forces to recapture the area.[59] In that instance, the minimalist approach ended up contributing to the violence. However, a gentle image is invaluable for militaries engaged in this type of conflict. They are embroiled in a war of perception and they draw legitimacy from endorsing the population-centric method. This is not to say that the aspirations toward restraint are false or misguided. They may well be genuine. Nevertheless, the level of violence is apt to exceed what is achievable with the best of intentions simply because of the complex assemblages of risk. In this environment, perception management is vital regardless of intentions. Security forces must consistently assert those intentions and construct their own narratives to impart meaning to events.

Simpson describes the war in Afghanistan as a quest for building the right narrative about the war for a series of audiences:

---

57. Thomas R. Mockaitis, *The COIN Conundrum: The Future of Counterinsurgency and US Land Power* (Carlisle Barracks, PA: US Army War College Strategic Studies Institute, 2016).

58. Daniel Branch, "Footprints in the Sand: British Colonial Counterinsurgency and the War in Iraq," *Politics & Society* 38, no. 1 (2010): 15–34; Douglas Porch, *Counterinsurgency: Exposing the Myths of the New Way of War* (Cambridge: Cambridge University Press, 2013).

59. Ledwidge, *Losing Small Wars.*

> Definition of the outcome of the Afghan conflict for the international coalition extends into the perceptions of audiences well beyond the insurgency. The Afghan people are deemed to be a central audience. Beyond Afghanistan the perception of the conflict's outcome within the Muslim world, and particularly in Pakistan, for instance, is a key factor. Moreover, the outcome in Afghanistan has global implications in terms of the credibility of the North Atlantic Treaty Organization (NATO), not least in terms of Chinese audiences.[60]

Of course, the audience can be subdivided still further into various constituencies within the host country, into partisan camps at home, into more and less amiable international audiences, etc. Here the logic of individuation can cause the taxonomy of audiences to become infinitely complex in the effort to identify specific grievances that can be addressed or identities that can be mobilized. Simpson says that this project of meaning making is unavoidably political, to such an extent that "the boundary between military and political activity is blurred."[61] So on the one hand perception management drifts inexorably into more focused messages, while on the other it must take on an expanding range of concerns until one narrative stretches across all aspects of the operations.

Within the population-centric framework, the indigenous civilians constitute the single most important audience. They are the human terrain. Conceptualizing these people as an audience radically transforms the norms that counterinsurgency forces must follow. If civilians are an audience, then some sense of moral legitimacy must be maintained to secure their support. This legitimacy can only come from providing security at a time of profound insecurity, but security efforts must stop short of inflicting new harms. They must not produce risk boomerangs. This constitutes a major shift in armed forces' orientation. Traditionally, militaries have been viewed as protectors of their states and representatives of their nations. They had fairly narrow constituencies, with few, if any, responsibilities to the citizens of third-party states and none at all to the citizens of opposing states. During the conventional wars of the nineteenth and twentieth century, enmity toward an opposing state was a generalized enmity against all citizens of that state. With population-centric counterinsurgency, foreign civilians become key constituents. They must be separated from the enemy belligerents and imbued with flexible civic identities that can be remade as needed.

---

60. Simpson, *War from the Ground Up*, 3.
61. Simpson, *War from the Ground Up*, 4.

Blame Shifting and Image Management

Counterinsurgency is shaped by institutional mediation at the military and political levels. Militarily, mediation is clearest from the outsourcing of security to local fighters. The host nation's military and police must be built or rebuilt and trained. Ideally they should take over a larger share of the security burden until the intervening forces are able to withdraw or revert to a purely advisory role. As the 2014 version of the US counterinsurgency manual states, "to be successful, counterinsurgency forces must work with and share risks with the host-nation forces and the population."[62] British and American personnel regularly work with local soldiers and police. Russian special operatives fight alongside Syrian soldiers and Ukrainian paramilitary forces. French soldiers train and support their counterparts in Mali. Fighting through these intermediaries reduces the risk of sustaining casualties and fosters perceptions of greater local control over the war effort. At the political level, the regime being supported by the intervening security forces plays a mediating role. Interveners strive to collaborate with local elites or to create new elites who can provide support. They are often kingmakers who set the political agenda by helping to design institutions and backing politicians who favor their goals. For example, Hamid Karzai's political career was dramatically accelerated by assistance from the intervening coalition, especially the decision to install him as the head of the interim government.[63] Support from the local government in turn gives the intervention legitimacy and facilitates blame sharing.

New technologies and practices of war, such as special operations missions, drones, and nonlethal weapons are justified (at least in part) with appeals to how they will reduce civilian casualties. Employing these more limited weapons shows an intent to improve normative compliance and it is often one that succeeds in reducing civilian casualties in practice. Nevertheless, policymakers routinely attempt to magnify the real drop in civilian casualties by concealing those civilians who are injured or killed.[64] They do not merely promise that drones will inflict fewer civilian casualties than gravity bombs; they promise a nearly perfect absence of civilian casualties. The promises exceed what even the most circumscribed attacks can achieve. However, institutional and physical mediation combine to provide more

62. *FM 3-24/MCWP 3-33/5 Insurgencies and Countering Insurgencies*, 1–11.

63. Steve Coll, *Directorate S: The CIA and America's Secret Wars in Afghanistan and Pakistan, 2001–2016* (New York: Penguin, 2018).

64. James Igoe Walsh and Marcus Schulzke, *The Ethics of Drone Strikes: Does Reducing the Cost of Conflict Encourage War?* (Carlisle Barracks, PA: Strategic Studies Institute, 2015).

opportunities for gatekeepers to control information. It is almost certain that the actual civilian casualties inflicted by drone strikes exceed the estimates offered by politicians. Nevertheless, assessments of these casualties differ wildly because of impediments hindering access to the necessary information. The narrower the range of people involved in these operations, the more it is possible to construct self-serving narratives.

This deception seems to be even more important for maintaining narratives of sensitivity toward civilians as the casualty numbers shrink. With fewer casualties, attention shifts onto the individuals who are harmed: individuals whose faces we can see and whose lives we can hear about. Following the logic of individuation, we know the names of specific people who were killed in drone strikes. Debate surrounds the justifications for killing Anwar al-Awlaki and Jihadi John. Journalists and aid workers write scathing rebukes of drone warfare that focus on Abdulrahman Anwar al-Awlaki. The ability to focus on individual victims increases the affective connection with those on the receiving end of attacks. It is easier to empathize with specific victims than abstract figures. Moreover, each reduction in civilian casualties highlights attacks against civilians as deviant. Whereas it was once possible to explain a stray gravity bomb striking a civilian home as an accident, the same excuse seems hollow when a strike is conducted using a smart bomb and proceeded by intensive surveillance that should be able to identify civilians who are at risk. Even a single unintended casualty evinces a mistake. And if there is only a single mistake, it seems all the more tragic because it is exceptional.

Conclusion

States have long had to contend with irregular fighters and have struggled to find more successful ways of defeating them. They have at times prevailed, but successes are rare and generally come at a high cost. States have routinely failed to defeat insurgencies, especially since the Second World War. Only a few exceptions like the British campaign in Malaya or Russia's subjugation of Chechnya ended with state military forces prevailing, while dozens of others have ended with an ignominious withdrawal from the contested area. Yet states are continually drawn into new campaigns against irregular fighters, and there is no end to proposals promising to finally provide them with the formula for success. The population-centric strategy stands out as being the most compelling at present. The United States and the United Kingdom explicitly strive to cultivate a local base of support in contested areas. In the wake of the Second Intifada, Israel has made renewed efforts to manage perceptions of its conflicts with the Palestinians and to enact more stringent restrictions on the treatment of civilians. Even Russia has taken steps toward showing greater restraint since its wars in Afghanistan and Chechnya, as its forces in Ukraine and Syria now operate on a lower scale and more covertly. Counterinsurgency in the twenty-first century is continuing to move in the direction of population-centric approaches. This trend is likely to continue so long as counterinsurgency remains the dominant mode of warfare, and other types of fighting will at least initially be undertaken by militaries designed for that mission.

The population-centric approach is explicitly premised on efforts to manage perceptions, especially those of civilians in the conflict area. Those people are transformed into a human terrain of fighting that supersedes geography. Appealing to the population depends on attempts to identify the interests of specific constituencies, and they are in turn carried out by soldiers who take on unprecedented autonomy as individual actors. In this context, soldiers become highly politicized, as does the military as a whole. The military mission is inseparable from political aims, which often have a partisan character, making it virtually impossible to maintain the military profession's distance from political disagreements. Strategic narratives display a reflexive uncertainty about the nature of the military profession and the complex risk tradeoffs that soldiers are forced to make. They are torn between competing demands that can never be perfectly reconciled.

8 • The Minimization of War

Degradation Supplants Decisive Battle

Of all the topics I discuss in this book, degradation is probably the one that receives the least attention in scholarly and media discussions. It is not associated with any particular weapon system, nor does it require high-profile organizational restructuring. However, it is an extremely important development that warrants scrutiny because of the role it has played in operations since the end of the Cold War. In fact, the comparatively low profile is part of the appeal. As I define it, degradation is the practice of attacking enemy forces to accomplish modest objectives, in ways that minimize the commitment of one's own forces, and without aiming to destroy the enemy. It is an attempt to achieve restrained victories that have limited impact, which runs contrary to the conventional war impulse of decisively crushing enemies to force unconditional surrender. This reduces the risk of unintended consequences.

Histories of war often convey a sense that war is among the chief driving forces of change, destroying empires, changing borders, overturning regimes, and driving expansion; they describe war as a revolutionary force. Degradation stands in stark opposition to this image. It is a deliberately conservative style of fighting; it is war to keep things as they are, or for incremental change. The political objectives of degradation tend to be small, calculated shifts, such as the removal of a leader or change of a specific policy. It is worth noting that this practice has been pioneered by Western states, which are often characterized as pursuing decisive warfare.[1] Endorsing this conservative strategy is a

1. For a detailed argument in favor of the view that Western warfare is characterized by decisive engagements, see Victor Davis Hanson, *The Western Way of War: Infantry Battle in Classical Greece* (New York: Random House, 2013). For more moderate evaluations that characterize decisive warfare as being aspirational or belonging to certain periods, see John Lynn, *Battle* (New York: Basic Books, 2003); Weigley, *The Age of Battles*; Brian Bond, *The Pursuit of Victory* (New York: Oxford University Press, 1998).

sign of those states' efforts to preserve the world they have created and to protect their place in it against rising powers, including nonstate actors.

The recent air campaigns against the Syrian government and the Islamic State provide some of the best examples of degradation in practice. As I write this, various Kurdish, Iraqi, and Syrian forces are fighting the Islamic State on the ground and attempting to seize its territory, yet the most powerful belligerents involved in the conflict—the United States, United Kingdom, France, Turkey, and Russia—are pursuing strategies of degradation. They launch missiles, drop bombs, support local proxies, and mount special operations raids with specific, limited objectives that minimize their exposure to risk. They lend support to one side of the Syrian civil war or another, but always with a sense of restraint. Their operations are designed to tip the balance of power toward the favored side, not to bring a decisive end to the war. Over the past three decades degradation has also been employed in places such as Libya, Iraq, Yemen, Somalia, and Kosovo. It is one of the preferred strategies for governments wishing to take military action—or at least wishing to appear to take military action—without bearing the burdens of protracted warfare and reconstruction. Degradation aims to achieve strategic and political outcomes with small uses of force on the tactical level. It envisions a handful of tactical decisions guided by strategic and political thinking that resonate back up, thereby collapsing the distinction between these levels of thinking. Western states still initiate some major conventional operations aimed at remaking entire regions, such as the 2003 invasion of Iraq, but these are exceptional. Lower-intensity conflict is usually preferred.

The exact form of degradation campaigns varies depending on the countries involved, their objectives, and the weapons available. Degradation is not reducible to a specific weapon system or attack type. It is instead an overall guiding rationale, a way of thinking about the use of force. It is a middle path existing somewhere between war (at least war in the conventional sense of a decisive clash of arms) and peace. It is war in a minimal sense. It is war without the appearance of war. It embodies a post-Westphalian sensibility that "war should not risk the stability of the national economy, society or political system."[2] One reason why this topic has not achieved the same prominence as others I discuss is that it tends to be a product of political, rather than military, priorities. Degradation is a logic invoked by heads of state more than heads of armed forces. This reflects the source of its appeal. Degradation becomes attractive when policymakers feel compelled to act—when the cir-

2. Martin Shaw, *The New Western Way of War: Risk-Transfer War and Its Crisis in Iraq* (Malden, MA: Polity Press, 2005), 73.

cumstances seem to demand some use of military force and when public opinion strongly favors it—but when a serious commitment would be too risky. The objectives are just as much to manage the perceptions surrounding use-of-force decisions and to manage global political responses as to achieve political outcomes within the area of operations.

The attempt to chart a risk-averse middle course in war is only plausible when armed forces possess asymmetric advantages over those who are being targeted. Precision-guided munitions, stealth, advanced communications technologies, and improvements in intelligence collection are technological prerequisites for finding and attacking specific enemy targets in pursuit of narrow goals. It is only when a stronger state can impose its will on a weaker adversary with technologically sophisticated weaponry that it becomes possible to coerce without much risk of escalation. This approach can look like conventional warfare because of the instruments involved, as in cases when American aircraft bombed Syrian military installations or when it launched cruise missile attacks against Saddam Hussein's palaces, but it is how these instruments are used that matters. By the same token, it is important to see that the asymmetry characteristic of degradation can exist during interstate conflicts, which is to say that it is not simply a method of fighting violent nonstate actors. Wars between states with vastly different military capacities can also take an asymmetric form.

I adopt the term "degradation" because this is the name most commonly employed by senior American leaders across multiple presidential administrations. They routinely describe their limited strikes as degrading the capacities of the current enemy, whether it is Iraq, the Islamic State, Syria, etc. The concept of degradation shares some characteristics with effects-based operations (EBOs). As one of the leading EBO proponents, David Deptula, says, "Well beyond the activity of destroying an opposing force lies the ultimate purpose of war—to compel a positive political outcome."[3] This goal of achieving political outcomes without destroying the enemy is something that EBO and degradation share. However, Deptula emphasizes that EBOs depend on launching parallel strikes against many different types of targets simultaneously, which suggests that attacks will usually employ overwhelming force. I use the term "degradation" to shift the focus to the use of small-scale attacks aimed at compelling positive political outcomes with minimal force in a single instance or that builds incrementally over time. In some cases, degradation may require major military operations, but this is atypical.

3. David A. Deptula, *Effects-Based Operations: Change in the Nature of Warfare* (Arlington, VA: Aerospace Education Foundation, 2001), 5.

The defining characteristic is not the scale of the violence but rather the attempt to employ precisely calibrated violence to achieve specific outcomes and with minimal risk of unanticipated consequences.

Indecisive War

The United States provides the archetype for this style of warfare and is my primary reference point. Degradation arose during the 1990s, at a time when the country was simultaneously consolidating its status as a world leader and reducing military expenditures. The Gulf War proved that the United States could still win conventional wars, but the fight was expensive and impossible to reproduce against every opponent. The United States needed a cost-effective way to manage risks at a time when it sought to play the leading role in protecting international security. The Gulf War reflected a nascent sense of the degradation approach insofar as ambitions were restricted to the liberation of Kuwait without regime change in Iraq. The goal of the war was essentially the conservative project of restoring the status quo with a limited military action short of forcing an unconditional surrender. The war likewise showcased some of the new technologies that would make this approach possible. Despite being used on a relatively limited scale, precision-guided munitions (PGMs) generated a disproportionate amount of media coverage and helped to popularize an imagined future in which the United States would be able to quickly and effectively kill its enemies without putting soldiers directly at risk.[4] Strategic thinking associated with the revolution in military affairs, and effects-based operations in particular, attempted to capitalize on the lessons learned from the Gulf War by emphasizing the possibilities of winning wars more easily and achieving political objectives more directly, with the help of precision and stealth technologies.

The costs facing American policymakers were not only financial and material. They were also perceptual. Fear that the American public would not support a protracted war after Vietnam urged moderation when committing troops. The perceived public backlash to casualties during operations in Somalia in 1993 gave new life to the fear of a casualty-averse American public limiting US military engagements.[5] There was a disconnect between the enormous power of the US armed forces and perceptions of what it could

4. Kahl, "In the Crossfire or the Crosshairs?"
5. See Luttwak, "Toward Post-Heroic Warfare."

achieve. The challenge was to use force without provoking backlash and while still presenting an image of American strength that could be useful in deterring potential adversaries. A further complication came from the goal of managing this military dilemma while legitimizing America's status of global hegemon by making it appear as a benevolent force resolving civil wars and humanitarian disasters.

In response to this challenge, the Clinton administration pursued a series of military ventures that established the degradation framework by limiting the risk to everyone involved in conflicts and carefully managing media narratives. US forces in Iraq imposed no-fly zones and a harsh sanctions regime to limit Saddam Hussein's resources. Throughout the decade a series of small air campaigns, such as Operation Desert Fox, checked the domestic use of Iraqi military power and punished Hussein for failing to comply with UN mandates. The attacks accomplished limited objectives, at limited cost, and with limited risk. Loss of life was less overt than in major conventional operations, which circumvented domestic opposition and supported the narrative of the United States as a benevolent hegemon. Operations in Bosnia and Kosovo throughout the 1990s further developed this mode of warfare. Both interventions were characterized by a heavy reliance on air power and an incremental escalation of hostilities, which indicate the goals of achieving peace with minimal risk to US forces and to local populations. Moreover, these operations helped to spread degradation practices to European allies, which faced similar pressures from domestic constituencies.[6]

The military value of restricted operations during the 1990s was often difficult to substantiate. The missiles launched during Operation Infinite Reach in 1998, for example, struck a pharmaceutical plant in Sudan, based on tenuous intelligence linking it to the manufacture of chemical weapons, and terrorist training camps that were largely abandoned. Strikes in Iraq likewise did little to seriously threaten Hussein's control or compel lasting compliance with UN mandates. Nevertheless, the strikes were politically useful. They made it possible to retaliate against terrorist organizations, punish dictators, and mount humanitarian interventions, all at a relatively low cost in terms of lives lost and money spent. Reliance on cruise missiles helped to entrench the imagined future of effective risk-free war in which it was no longer necessary to deploy soldiers in foreign missions, but further innovation was needed to make this kind of ranged approach more effective in reaching the right tar-

---

6. David Halberstam, *War in a Time of Peace: Bush, Clinton, and the Generals* (New York: Simon & Schuster, 2001); James D. Boys, *Clinton's Grand Strategy: US Foreign Policy in a Post-Cold War World* (London: Bloomsbury, 2015).

gets. Here it is important to note that the political utility of degradation is not measured primarily by the direct outcome in foreign states (as in the case of effects-based operations) but by those effects combined with domestic and international political considerations. As Usha Sahay points out, the goal of Infinite Reach went beyond killing bin Laden to "communicating a message of strength and resolve."[7] In that case, appearing tough on terrorism with little cost was politically expedient even though it had limited impact on al Qaeda.

The United States has continued to follow a degradation strategy in most operations during the twenty-first century. The war in Afghanistan was initially pursued as a small venture with minimal investment and heavy reliance on local forces, only growing slowly after the failure to capture Osama bin Laden. Operations in Syria, Libya, Mali, and Yemen under the Obama and Trump administrations were likewise focused on reducing enemy military effectiveness. In most of these cases, the goal was to degrade the enemy's capacity from a distance as a way of assisting allies fighting on the ground. The 2003 invasion of Iraq was an exception that proves the trend. What began as a conventional war with sweeping ambitions of regime change and building a sense of national identity proved far more costly than Bush administration officials predicted, and it now serves as a cautionary tale urging limited investment in other conflicts.[8] Major conventional wars may therefore continue even in an era in which degradation is more prevalent, but they are likely to be less common and less attractive as models of good strategy.

Degradation is most closely associated with states that have a stake in maintaining the political status quo, either on a global or regional level. These countries have little to gain from pursuing revolutionary change and much to lose, which makes risk aversion sensible. Not coincidentally, these are the same states with the technological capacity to use force in the way this approach calls for. Asymmetric advantages permit them to fight with little risk of escalation or sustaining casualties; they have the power to choose when they fight and the freedom to decline wars that could be too costly. Degradation is particularly appealing for global hegemons guarding their status, which is one reason it is one of the preferred approaches for the United States. Hegemons cannot hope to intervene everywhere. They must maintain stability cheaply and efficiently and while keeping their options

7. Usha Sahay, "Syria, Signaling, and Operation Infinite Reach," *War on the Rocks* (September 2013), https://warontherocks.com/2013/09/syria-signaling-and-operation-infinite-reach/

8. Anthony H. Cordesman, *The Iraq War: Strategy, Tactics, and Military Lessons* (Westport, CT: Greenwood, 2003).

open for the next emergency. Whenever possible, the specific operations can be augmented with help from allies or private contractors, which reduces the risk borne by any single partner and helps to spread responsibility for managing unintended consequences.

Degradation is also advantageous when attempting to build coalitions, which supports the US preference for assembling and leading multilateral forces. Successful coalition warfare depends on building a sense of mutual interest and on limiting the collective action problem.[9] Both of these goals can be advanced by minimizing the extent of a conflict via the degradation logic. The more modest the strategic and political goals are, the easier it is to establish a collective interest in achieving them without alienating any partners. A broad array of countries can join the effort to punish Bashar al-Assad's use of chemical weapons, for example, even though they have much different priorities in the region. More sweeping changes like those of the Bush administration during its occupation of Iraq are antithetical to collective agreement, and hence pose more problems for coalition building. Low expectations in terms of military commitment also make it possible to participate without much investment of resources or risk of provoking domestic political opposition, mitigating the collective action problem. Partners may therefore contribute to the war effort with relatively small gestures while still being assured of their importance to the larger project.

Finally, degradation is a strategy for regional hegemons and states attempting to enforce stability in regions where they have a preponderance of military force. Israel stands out as a clear example of this. It has considerable freedom of action when operating in the western half of the subcontinent, but it must also project a positive image to the rest of the world, especially the United States. Its strike against an Iraqi nuclear reactor in 1981 was an early instance of the country experimenting with degradation with a small act of violence to achieve limited objectives. Following several costly and inconclusive counterinsurgency campaigns in Lebanon and the Palestinian territories throughout the 1980s and 1990s, it adopted degradation as its preferred strategy. Its operations against the Palestinian territories are now generally characterized by small strikes that aim at destroying infrastructure and tunnels. Its participation in the Syrian civil war has likewise been characterized by a minimal use of air and ground power to protect its border and contest Iranian influence in the country.[10] This allows it to keep local rivals in

---

9. Scott Wolford, *The Politics of Military Coalitions* (Cambridge: Cambridge University Press, 2015).

10. Ahron Bregman, *Israel's Wars: A History since 1947* (London: Routledge, 2016).

check while minimizing the risks of inciting larger conflicts or suffering from serious international repercussions.

The Risks of Victory

Degradation is essentially a form of risk management. It reflects a desire to launch military interventions with minimal exposure tactically, operationally, strategically, and politically. On a tactical level, the overriding goal when degrading enemies is to destroy key targets, such as factories, power stations, transportation networks, and leaders without excessive violence or unanticipated consequences, like civilian casualties or mass devastation of industry. The targets have political and strategic significance that reaches beyond the tactical level, as they are selected to provide large payoffs that outweigh the chances being taken during the strikes. It is important to note that the targets need not always be high-profile leaders or government facilities; this is why degradation is not identical to targeted killing. Any action in which a few select targets are expected to yield major political benefits can qualify. In the struggle against the Islamic State, President Obama explicitly invoked the logic of degradation when US forces targeted captured Iraqi military hardware. In that instance, the reduction of enemy strength was designed to give local allies an advantage on the ground. Coalition attacks in Libya in 2011 were likewise directed at a range of military targets whose destruction would facilitate an insurgent victory. Enemy leaders were among the targets selected, but the campaigns were not exclusively concerned with decapitation.

Although degrading attacks may vary in magnitude and scope, they always require a high degree of precision, which is why this practice is relatively new and coincides with the advent of more precise weaponry. Harm must be confined to the immediate target area. Lowering the extent of damage and minimizing loss of life are key considerations when avoiding unintended consequences and escaping responsibility for reconstruction. Narrow targeting reflects a morally admirable effort to limit the effects of violence. Civilians outside government are placed off-limits, in accordance with international law, and so too are many enemy soldiers who are tangential to the main effort. Degradation is therefore compliant with the just war principles of discrimination and proportionality. At the same time, the morally commendable restriction on the scope of violence comes with the politically and morally undesirable effect of lowering the barriers against launching low-intensity coercive operations. Control of risks is essential for making war appear to be a legitimate instrument of coercion, but it is possible for this legitimacy to be abused when it covers up morally objectionable operations or operations whose consequences have not been adequately considered.[11] In other words, cleaning up the appearance of war erodes the conceptual barrier between war and peace. The result is evident in ongoing US operations in the Middle East, with the country perpetually involved in small attacks that do not amount to conventional warfare.

There is also a broader limiting of harm in the sense of preventing major disruptions that could cause risk boomerangs. Degradation is a gesture toward moderation and conservation. It is based on leaving the world largely as it is, such as by reducing an enemy's capacities for aggressive action without eliminating the enemy. Completely destroying enemies may introduce a host of new problems for the victor: governing the conflict area, restoring security, and repairing the infrastructure. These require long-term investment and are politically dangerous. The extent of these costs is clear from the wars in Iraq and Afghanistan. In both cases the American and British militaries were able to achieve quick and decisive victories only to find that this gave them an uncomfortable degree of responsibility for postwar reconstruction and security. This task proved to be far more difficult than many had anticipated, and it undermined the Bush and Blair administrations' claims that overwhelming popular support and desire for democracy would guarantee stability.[12] The struggle in Iraq was severely aggravated by the de-Baathification program: a decision to remove all Ba'ath Party members down to the fourth level of the organization's hierarchy and to disband what was left of Iraq's military. This decision tore down the government and security apparatus with the goal of rebuilding it. When the reconstruction process turned out to be too difficult for the overstretched coalition forces—especially with the insurgency being fueled by former soldiers—it was impossible to look to the defunct institutions for assistance.

If the overall goal of a campaign is to see that the enemy is wiped out, the logic of degradation dictates that the final destruction must be affected by a local force fighting on the ground. For example, the European militaries intervening in Libya destroyed Kaddafi's palaces and military installations to even the odds between his forces and the rebels, thereby allowing the rebels

11. Martin Shaw, "Risk-transfer Militarism, Small Massacres and the Historic Legitimacy of War," *International Relations* 16, no. 3 (2002): 343–60.

12. David L. Phillips, *Losing Iraq: Inside the Postwar Reconstruction Fiasco* (New York: Basic Books, 2009); Dov S. Zakheim, *A Vulcan's Tale: How the Bush Administration Mismanaged the Reconstruction of Afghanistan* (Washington, DC: Brookings Institution Press, 2011).

to triumph.[13] Without ground forces committed and with responsibility for the conflict shared between multiple actors, the intervening states had considerable freedom in determining whether and how to respond to the resulting instability. That left the Libyans in control of the country and gave them the work of establishing the postwar government. The intervention against the Islamic State has followed roughly the same model. The United States and its allies have sought to reduce that organization's military power, while leaving the burden of ground combat and reconstruction with Kurdish, Iraqi, and Syrian forces. This is another goal that is in some ways admirable, as it promotes local autonomy and avoids the problem of humanitarian interventions being used as a pretext for occupation and economic exploitation. However, such interventions have the undesirable outcome of leaving local actors with a reconstruction task that may exceed their capacity.

Constructing the Best Impression

Risk management is also a matter of reducing the likelihood of sustaining military casualties during operations, a goal that is advanced when strikes are limited to essential targets and carried out using weapons systems that establish risk asymmetries. Casualty sensitivity is more of a cost-benefit analysis than a fixed condition. John Mueller and Bruce Jentleson have shown that people make decisions about when military force is needed and how many casualties are permissible with an eye to the policy objectives and the benefits these may have.[14] Studies of casualty sensitivity have also shown that expectations of success play a larger role in directing attitudes toward conflicts than the number of military casualties sustained.[15] The civilian public is generally prepared to take some losses when these are sustained in pursuit of important policy objectives and success appears to be likely. However, these two conditions are rarely met in post–Cold War conflicts. Wars during this era are often about defense of other states or defense of threatened populations, goals that tend to receive lower levels of support

13. Christopher S. Chivvis, *Toppling Qaddafi: Libya and the Limits of Liberal Intervention* (Cambridge: Cambridge University Press, 2013).

14. John E. Mueller, *War, Presidents and Public Opinion* (New York: Wiley, 1973); John Mueller, *Policy and Opinion in the Gulf War* (Chicago: University of Chicago Press, 1994); Bruce W. Jentleson, "The Pretty Prudent Public: Post Post-Vietnam American Opinion on the Use of Military Force," *International Studies Quarterly* 36, no. 1 (1992): 49–73.

15. Christopher Gelpi, Peter D. Feaver, and Jason Reifler, "Success Matters: Casualty Sensitivity and the War in Iraq," *International Security* 30, no. 3 (2006): 7–46.

than wars of self-defense.[16] When pursuing goals that are perceived as being less salient, there is a higher chance of public backlash if soldiers are injured or killed. Anticipations of success are likewise difficult to manage if wars drag on for decades without reaching any clear end point, which forces the public to rely on casualties as the primary metric for evaluating progress during wars.

When the policy objectives have low salience and victory is illusive, policymakers must minimize the risk of sustaining military casualties to retain public support for interventions. This makes the risk-hedging practice of degradation extremely attractive. However, there is an ironic effect of this effort to evade risk. Degrading enemies rather than seeking a decisive confrontation extends the length of wars and reduces the likelihood of a clear success at the end, further increasing the salience of casualties. Such a minimal commitment to wars likewise suggests that fighting is not essential for national security. And each time the degradation approach is used, the public is conditioned to expect that it is possible to wage wars with minimal risk. These processes join together to create a trap for risk-averse leaders. Evading risk by adopting the strategy of degradation entrenches them in this approach and urges them toward ever more restricted attacks.

Perception management is vital to counteracting the public opinion costs of waging low-salience wars that have low or distant prospects of victory. Before hostilities commence, the public needs reassurance that the fighting will be asymmetric and that allied forces will be relatively safe. Civilian audiences were given this confidence in advanced weaponry and the revolution in military affairs (RMA) during the 1990s with impressive videos of cruise missiles and smart bombs perfectly striking their targets. The weapons seemed revolutionary and, with their capacity to hit targets without putting Americans at risk, encouraged feelings of casualty aversion by making the loss of life seem unnecessary. The proliferation of these weapons to other countries has extended the enthusiasm for high-tech solutions beyond the United States, especially to America's allies in Europe and Israel. Building this sense that new technology can achieve victory at low costs requires sustained media efforts, including spectacles of these new technologies in action in news broadcasts, documentaries, movies, and video games. News and entertainment media jointly build confidence that can limit opposition to new military ventures.[17]

---

16. Bruce W. Jentleson and Rebecca L. Britton, "Still Pretty Prudent: Post–Cold War American Public Opinion on the Use of Military Force," *Journal of Conflict Resolution* 42, no. 4 (1998): 395–417.

17. Roger Stahl, *Militainment, Inc.: War, Media, and Popular Culture* (New York: Routledge, 2009); James Der Derian, *Virtuous War: Mapping the Military-Industrial-Media-*

During war, confidence in these technologies and in the overall strategy of degradation is maintained by positive imagery of the technologies and efforts to control the appearance of casualties. In many cases, allied, civilian, and enemy casualties are hidden outright via media censorship and the control of information.[18] In other cases casualties are selectively shown when they can be fit into narratives that advance the war effort, such as when Syrian government atrocities against civilians are used to demonstrate the importance of launching more attacks.[19] It is no accident that an array of commentators have decried recent operations characterized by degradation as "risk-free war" or as "spectator sport warfare."[20] When war is remade along the lines of degradation, eschewing decisive pitched battles and concealing violence, it feels less like war. The derealization of war is a natural byproduct of violence that is calibrated to prevent disruptions of the international system. Ignatieff aptly observes of the intervention in Kosovo that "It was the kind of war fought by peoples who have known fifty years of peace; the kind of war a nation fights when it wants to, not when it must, when values rather than survival are on the line; when commitment is intense but also shallow."[21]

Politically, the increased emphasis on deploying precision weapons and other "usable weapons" to launch attacks with relatively limited goals has costs and benefits. On the positive side, it helps states minimize the chance of sustaining casualties, thereby allaying the fears that a casualty-averse public will react negatively to operations and consider them too costly to be sustained. This raises the prospect of making it easier to continue fighting for longer periods of time and facilitating "wars of choice" in which there are no compelling national interests that would help to build enthusiasm for the war. The cost of degradation as a strategy is that it introduces new grounds for criticism. In particular, it exposes leaders to charges of weakness and indecisiveness. Both critiques were deployed against Presidents Clinton and Obama, who were among the most prolific advocates of this approach.

Entertainment Network (New York: Routledge, 2009); Katy Parry, "The First 'Clean' War? Visually Framing Civilian Casualties in the British Press During the 2003 Iraq Invasion," *Journal of War & Cultural Studies* 5, no. 2 (2012): 173–87.

18. Parry, "The First 'Clean' War?"

19. Marcus Schulzke, "Concealing or Confronting the Costs of War: Modes of Representing Casualties in American News and Entertainment Media," *Critical Studies on Security* (forthcoming).

20. Colin McInnes, *Spectator Sport War: The West and Contemporary Conflict* (Boulder: Lynne Rienner, 2002).

21. Michael Ignatieff, *Virtual War: Kosovo and Beyond* (New York: Picador, 2001), 4.

Minimization through Mediation

Efforts to control audience perceptions of conflict through skillful reporting and narrative building depend on mediation. Mediation provides opportunities to select information and frame it advantageously. The more layers information must pass through, the more opportunities there are for control of information. At the same time, information control is less obvious when audiences feel that they have direct access to events. Digital technologies have the dual benefits of offering new opportunities for information control and providing a heightened sense of authenticity. This is evident from the images captured by precision-guided munitions and drones. These give audiences a clear view of the battlefield and therefore promote a sense of audience engagement in events. However, military and political leaders are free to select those images that best fit the narrative they want to promote. The result is that those images seeming to provide direct access to events on the ground do so while also shifting attention away from counterevidence and building confidence in the official narrative of events. Thus, the US military may release videos of successful strikes against Islamic State leaders but sanitize information about accidental attacks on civilians that conflict with narratives of precision warfare. Evidence of the latter attacks tend to come from third parties offering counternarratives rather than from belligerents themselves.

Attacks are mediated not only through the digital weapons that are involved but also in a more abstract sense by the effort to share the burden of intervention. A state acting alone risks being seen as an aggressor that is unilaterally disrupting peace or escalating an ongoing conflict through military intervention. It must also commit more resources because it must sustain the fight without being able to count on external support. Acting alone would therefore put a state in a dangerous position both when attempting to control perceptions of the intervention and when attempting to manage risks. It becomes more difficult to generate a positive image of military action when it is being pursued by a single state rather than a coalition of partners motivated by a common goal. Coalitions grant moral authority and, when they are powerful, a sense that they will inevitably triumph. This is the case in the Syrian intervention, which has arrayed some of the world's most powerful militaries in what has been cast as a moral crusade to destroy an evil organization. Spreading the burdens of fighting facilitates risk management in several ways. First, it allows belligerents to keep some resources in reserve to guard against threats elsewhere. Second, shared responsibility for the outcome of a conflict prevents any one state from being blamed, either for negative outcomes or for missed opportunities. Finally, any state suffering some reversal of fortune, such as losing an aircraft over enemy territory, can call on support from coalition partners.

In practice, the goals of mediation through coalition partners and how these are pursued depends on the relative power of the state involved. States make much different demands on their allies and mobilize these support networks in different ways. The United States tends to place relatively little weight on the actual military commitment from allies, instead seeking moral support that will help to confer legitimacy and manage perceptions. It routinely attempts to build coalitions to wage wars even when the other states involved are unwilling to commit military resources or lack the capacity to do so. Other states depend on mediation to augment their military capabilities. For example, France relied on an array of partners from around the world to provide soldiers, equipment, intelligence, and logistical support for its operations in Mali in 2012. This assistance helped France reduce its own commitment in the theater, thereby giving the intervention the minimalism of the degradation approach rather than elevating it to something more akin to a full-scale conventional operation.

Local allies are especially important; they confer legitimacy and fighting power. Degradation is not a strategy of self-defense but rather one of defending another state or group of people, as wars of self-defense tend to demand major conventional operations. The legitimacy of operations therefore rests heavily on whether the country or group that is ostensibly being defended offers support. The Syrian civil war illustrates this. Countries intervening against the Islamic State may lend their support to an array of different actors: the Syrian government, other Syrian rebel groups, Turkey, Israel, or Iraqi militias aligned with Kurdish, Sunni, or Shiite factions. The choice of allies partially depends on long-standing alignments. There is, of course, little chance that NATO countries would align themselves with the Syrian government or that Russia would align with the Kurds. Nevertheless, there are also more immediate interests at stake. Interveners benefit from engaging in degradation that helps participants who are sympathetic and capable of effectively prosecuting the war. The interveners are unlikely to support factions that will be unpopular for domestic audiences or those that have little hope of success.

When thinking about the partners within the conflict area, it is important to remember that degradation is a method of fighting that is employed by certain belligerents, not a characteristic of the war as a whole. Fighters who live in Syria and Iraq are engaged in a total war with existential repercussions. It is only the outside powers who can choose their level of involvement that have the opportunity to fight according to the logic of degradation. Degradation is therefore a practice reserved for strong states that bear little direct risk and a lower stake in the ultimate outcome of the fighting. The minimal operations they undertake are not meant to have a decisive impact in themselves, and they usually seek to uphold the status quo or something similar to it. Local allies are apt to have more ambitious goals in mind and are usually seeking a decisive outcome. Thus such wars can have vastly different meaning for the participants, even when those participants are fighting on the same side. This may require states engaged in degradation to manage their local allies by reining in ambitious plans that would upset the status quo. When this fails, the local allies being supported can generate new risks with victories that perpetuate instability in the region.

Contradictory Geographies

Degradation redefines the territorial dimension of conflict in several senses. First, it is informed by an aversion to ground warfare. Deploying ground forces signals a long-term commitment to capture and control an area. It also demonstrates a willingness to sustain military casualties when pursuing this goal. Degradation is guided by weaker commitments: a desire to fight without loss or the risk of protracted involvement. The clearest indication that a state is adopting the degradation strategy is the refusal to commit ground forces and a heavy reliance on air power or guided missiles. To the extent that ground units are deployed, these tend to be small numbers of special operations personnel to coordinate air strikes and strengthen local allies. Soldiers are not deployed in the numbers needed for seizing and holding territory. In most cases, they are not even meant to operate independently. They are instead part of integrated teams made up primarily of local allies who are expected to do most of the fighting and to suffer most of the casualties.

Second, operations are not limited to a particular space. Geographical boundaries are perceived not as sharp limits that operations cannot exceed but instead as hurdles to be overcome with the help of new technologies and organizational structures. Here degradation draws on some of the same strategic thinking that is evident in effects-based operations, although as I have established, it aspires to more limited uses of force with broader political influence. In a discussion of parallel or simultaneous strikes characteristic of EBO, Deptula argues that attacks should aim to transcend geographical and temporal boundaries. "Simultaneous application of force (time) across each level of war uninhibited by geography (space) describes the conduct of parallel warfare."[22] Degradation is part of a collection of overlapping strategic discourses that rely on similar technological capacities and that envision the supersession of terrestrial constraints. Whereas military strategists tend to favor operating concepts like EBO and the RMA, politicians invoke the language of degradation, which treats transgeographical military operations instrumentally.

Third, whereas many wars throughout recorded history have aimed at territorial conquest or the redrawing of borders, degradation aims to minimize shifts in the balance of power. It is a territorially conservative orientation. Borders must be preserved; power vacuums must be avoided. Managing conflicts entails keeping them confined to a particular territory—usually delineated by existing state borders—even though military operations themselves may range far beyond. It is only by ensuring that enemy efforts remain localized that the effects of the fighting can be contained. Shaw argues that "Precision defines war, narrows it down, and makes it a discrete and manageable business,"[23] and this same point can be extended to the strategy of degradation, which relies heavily on the logic of precision. Degrading enemies acknowledges the inevitability of conflict in a particular area while still striving to prevent it from having any major influence on regional politics. This makes it particularly attractive in the Middle East and North Africa, where ethnic and religious affiliations pose the danger of drawing more parties into a conflict and producing major shifts in the regional balance of power.

Finally, degradation makes it possible for states to pursue contrary goals simultaneously, and even to become embroiled in conflicting allegiances depending on the area of operations. Because the goals are territorially conservative, it is possible to find common ground with an ideologically diverse range of allies. There is likewise no need for a comprehensive global strategy, such as consistently supporting left-wing or right-wing governments. Thus, the United States may fight Iranian-backed Houthi militias in Yemen while still seeing Iran as an asset in the effort to defeat the Islamic State. This provides an interest in hindering Iranian support for Yemen (as well as for the Syrian government) while facilitating Iranian-supported militias' attacks against the Islamic State. Such a mixture of objectives is unthinkable within a consistent global perspective on war, yet it becomes entirely possible when each conflict is deliberately pursued as a self-contained effort alongside other self-contained efforts.

22. Deptula, *Effects-Based Operations*, 5.

23. Shaw, *The New Western Way of War*, 88.

A paradox should be evident at this point. States involved in degradation campaigns routinely violate the sanctity of sovereign spaces while also attempting to reassert local sovereignty and confine enemy efforts to a particular area. For example, operations against the Taliban and al Qaeda spill into neighboring countries, but the focus of nation building is rehabilitating a Westphalian state within Afghanistan, without redrawing the borders. States pursuing this approach are therefore guilty of a performative contradiction. Their conservative goal of preserving existing borders and defending/building Westphalian states is tacitly undercut by a method of warfare that shows little respect for sovereign authority or the sanctity of international borders. Such a double standard makes sense for the kinds of global and regional hegemons that are attracted to degradation. Their power depends heavily on acting beyond their own borders while preventing rivals from doing the same. Nevertheless, this practice cannot help but undermine the normative status of borders. The insurgents and weak states on the receiving end of asymmetric strikes, as well as third-party critics around the world, routinely decry the hypocrisy of military interventions.

As I have emphasized throughout the book, current trends in armed conflict point to the transformation of geographical barriers and the emergence of supra-territorial fighting that can spread rapidly beyond state boundaries. It should come as no surprise that attempts at containment ultimately fail when they resist this broader trend, and especially when degradation depends on many of the weapons, organizational structures, and practices that are fueling the transcendence of geographical constraints. The erosion of borders, at least as limits on conflict, will make degradation more attractive, perhaps even necessary. States will be virtually forced to pursue enemies beyond borders, especially when fighting insurgents, or risk being defeated.

The struggle against the Islamic State is one of the greatest efforts at degradation to date, involving a broad coalition of Western and Middle Eastern armed forces. It even brought rivals and enemies together in an unstable alliance held together by a mutual interest in keeping the Islamic State in check. And yet even with this unprecedented effort at containment, the war spread far beyond Syria's borders. The US-led coalition attempted to counteract the Islamic State's effort to redraw borders by defending Iraq against the encroachment. At the same time, the coalition launched attacks in Iraq and Syria (as well as in other countries where affiliates cropped up), essentially treating the Islamic State as the transgeographical entity it aspires to be. To add further complexity, the coalition partnered with local actors pursuing contrary goals. In some cases these actors even hoped for their own redrawing of boundaries, which coalition forces seem unlikely to honor. The Kurds, for example, are long-time American allies who receive protection and seek greater autonomy but are unlikely to convince the United States to support the creation of an independent state.

The Israeli strategy in Gaza can be characterized as degradation because efforts are made to confine Palestinian threats to specific, precisely bounded areas through the use of asymmetric force transcending those boundaries with the goal of managing, rather than decisively beating, the opposing forces. Those operations show two additional territorial challenges for degradation campaigns. First, despite cordoning Gaza with concrete walls and subjecting it to constant surveillance from the sea, air, and land, the tiny strip of desert has remained connected to the outside world by tunnels linking it to Egypt. Lacking control over the exit points, Israeli forces strive to collapse the tunnels with aerial bombardments or to take control of the outlets within Gaza itself. Once again there is the ironic effect of state borders becoming a hindrance to states as they attempt to protect the territorial status quo, and the efforts to seize control of the tunnels in Gaza force Israel to deviate from the strategy of degradation to counterinsurgency. Second, Israel struggles with territorial containment because news of its attacks spread internationally. Its recent campaigns have provoked international backlash, being condemned by news agencies, civil society organizations, and members of foreign governments. Here the deterritorialization of war is coupled with the growing importance of perceptions and perception management. The impossibility of containing information within the geographical space of conflict has the effect of extending the ideological battlefield to the global level.

One Step Forward, One Step Back

The strategy of degradation both affirms and undermines existing norms relating to the use of force. The strategy is heavily informed by the principles of proportionality and discrimination. A sense of proportionality directs the overall intent of the strikes and the means that are used to achieve them. With modest objectives that are concerned primarily with hindering the enemy's ability to employ military force, intervention can be minimized. There is usually no need for destroying the enemy government or rendering it incapable of functioning. Doing so would contradict this strategy and the advantages it offers. The 2003 Iraq War underscored this. The de-Baathification program went against the sense of proportionality that many within the US and allied militaries had developed over the preceding decade, and it ultimately lent credence to proportionality because of the ultimate failure of nation building. With minimal goals in mind, the level of force can be restricted to what is essential for destroying key military targets. This typically includes enemy leaders and key military infrastructure, which means that lower-level enemy personnel and small concentrations of enemy forces may escape destruction. One of the marks of states pursuing degradation is that they use some of the least powerful weapons at their disposal—Hellfire and missiles, smart bombs, and special operations forces—rather than armored units and large gravity bombs.

Degrading strikes are guided by the principle of discrimination insofar as they are meant to destroy military targets and care is often used to engage the targets when civilian casualties can be minimized. Avoiding civilian casualties is essential for maintaining the image of precision that guides the use of air power and special operations forces. In practice, missiles sometimes hit the wrong targets. Civilians are injured and killed. The roads and power plants they depend on are damaged or destroyed. However, the civilian victims are rarely the intended targets. They are not attacked en masse, as they were during the major conventional wars of the twentieth century. Their suffering is more the result of negligence and error than policies of deliberate civilian victimization. Belligerents are justifiably criticized for such errors and should pay the price for their mistakes, but the benevolent intentions matter. For one thing, the intention to avoid civilian casualties lends legitimacy to the principle of discrimination and to the just war tradition more broadly. On the other hand, firm commitment to norms of war underwrites belligerents' claims to the moral high ground. It transforms norms into weapons that can be wielded in the struggle to control narratives that impart meaning onto conflicts.

Degradation shows heightened efforts to conform to some norms of just war, especially the *jus in bello* guidelines that govern the use of force once a war begins. Ironically, conformity with norms during war is coupled with reduced interest in *jus ad bellum* norms relating to whether a war itself is justified. When states launch degrading strikes, they often attempt to present these as falling short of real war. The attacks involve military force, but are described as operations for "peacekeeping," "security," "humanitarianism," or some other legitimizing mission that can be defined as being outside the remit of what audiences would typically understand as war. Without entering real wars, states have greater freedom to use military force without facing normative and bureaucratic restrictions designed to avert armed conflict. Foremost among these is the principle of last resort, which requires states to exhaust all peaceful means of conflict resolution before using military force. If degradation is something other than war—a use of force short of real war—then it becomes normatively obligatory as a lethal option that states can invoke without ceding their moral authority.

Just war theorists continue to debate whether the kinds of limited strikes that I describe as degradation constitute a form of war or some new normative domain. They disagree about whether all *jus ad bellum* criteria need to be met when a state launches limited strikes or whether norms should be relaxed so states have greater scope for protecting their security without resorting to conventional warfare. This reflects a dilemma about how best to restrict wars. Is it better to continue affirming a high threshold for war at the risk of allowing the conditions that produce large wars to build up uncontrolled? Or is it better to authorize the frequent use of minimal attacks in the hope that these may be able to prevent major wars from occurring? Further complicating this issue is that degradation has mixed results. In some instances, such as the strikes against al Qaeda under the Clinton administration and the support for the Libyan civil war, the attacks appear to have failed. Al Qaeda's ability to operate did not suffer and the efforts to help Libyan rebels only assisted them in throwing the country into anarchy. The attacks on the Islamic State appear to have been more successful. Although the ultimate outcome of these remains unclear, it at least seems plausible to say that these helped Iraq defend itself against a serious threat. It is difficult to say whether degradation is justifiable as a means of averting war without more evidence of whether it truly does help to avert war.

Once again, the lack of territorial objectives complicates efforts to reach normative clarity. As I discussed in previous chapters, the norms of *jus ad bellum* are highly sensitive to territorial claims, especially when it comes to the principle of just cause. Because degradation does not aspire to territorial control, it is difficult to determine exactly when the use of force qualifies as war. A state may have clear cause for war when it faces invasion from a powerful neighbor, but it is more difficult to justify war when the more powerful state only launches a few limited strikes within its neighbor's territory without attempting to claim land. Without such a clear line between war and peace as an international border that must be physically crossed, the normative line between war and peace becomes ethereal. It has already vanished to such an extent that it is difficult to reach a consensus about when wars are being fought and what the objectives are.

Jus post bellum received relatively little attention in just war theory until the twenty-first century.[24] Victory in war was usually an opportunity to impose punitive measures and to gain at the loser's expense. Reconstruction programs to help the defeated country were rare, with the notable exception

24. See Brian Orend, *War and International Justice: A Kantian Perspective* (Waterloo, ON: Wilfrid Laurier University Press, 2000); Brian Orend, "Jus Post Bellum: The Perspective of a Just-War Theorist," *Leiden Journal of International Law* 20, no. 3 (2007): 571–91.

of assistance following the Second World War. Counterinsurgency, counterterrorism, and humanitarian relief have prompted unprecedented attention to fostering long-term security in conflict areas. Moreover, the widespread aversion to unnecessary war has built a sense that conflict must be redeemed by postwar rebuilding, which is a prominent theme in the recent *jus post bellum* literature. Unfortunately, postwar rebuilding is expensive, time-consuming, and difficult. Commentators have yet to agree on a particular model that this should follow, and it is doubtful that consensus will be forthcoming. Even specific principles of *jus post bellum* akin to those of *jus ad bellum* and *jus in bello* cannot be agreed on, which means that even well-intentioned reconstruction programs will lack a strong normative foundation. The normative challenges and confusion about effective approaches means that postwar reconstruction is an expensive, thankless, and often wasteful enterprise with little appeal. Degradation is an ideal strategy for prosecuting a conflict without venturing into postwar reconstruction. If the enemy is only weakened, then the resulting power vacuum will be smaller. If critical industries are damaged but not destroyed, they may be able to recuperate without substantial outside investment. Above all, a minimal commitment of ground forces helps to preserve some distance from the challenges of reconstruction. Thus states have a strong incentive to continue the practice of degradation as a way of minimizing normative, political, and financial risks.

Individuating Targets and Audiences

Degradation encompasses many of the innovations I discussed in previous chapters. There is an especially strong affinity with targeted killing and the methods of selectively attacking enemy leaders, drones, and special operations forces. Attacks on leaders are carried out with the hope that decapitating organizations will render them incapable of functioning, without necessitating conventional operations. However, whereas the individuation is manifest in targeted killing through the focus of attacks against enemy leaders, the individuation manifest in degradation extends to the material infrastructure. The attacks on Saddam Hussein and his palaces throughout the 1990s are a prime example of this. The goal was not simply to kill a particular person but also to impose punishments that would affect Hussein if the attacks failed to kill him. Similarly, attacks against the Islamic State use the techniques of targeted killing against material targets like armored vehicles and weapons production facilities. There is an individuation of things that mirrors the individuation of people in targeted killings. Thus targeted killing is routinely incorporated into degradation campaigns, but those campaigns also pursue more expansive objectives that go beyond the death or capture of key leaders.

When attacks are designed to even the odds so that local allies can prevail over their rivals, or when they are framed as punitive strikes, structures and vehicles can take on as much importance as human targets. The United States and its allies adopted a pattern of striking specific material objectives during their interventions in Iraq and Syria to produce controlled outcomes. For example, air strikes took out oil refineries to cripple the Islamic State's production, armored vehicles captured from the Iraqi government to even the odds between the Islamic State and the actors opposing it, and Syrian government runways to punish the Assad regime for using chemical weapons. These and other attacks were not meant to achieve a decisive outcome but instead to nudge the conflict in directions that were mutually beneficial to the coalition partners. They were, therefore, individuated in terms of the precise target selection and in terms of the narrow, calculated outcomes.

Degradation also promotes individuation among civilian audiences watching wars from a distance. As Christopher Kutz points out, conceptions of state sovereignty that emerged during the late sixteenth century defined war as a relationship between states, removing citizens (or subjects) from a direct stake in the fighting. "The fact that the sovereign was at war with another state would not mean that his subjects were at war with the other state. War could not be, in moral terms, a relation between the soldiers actually doing the fighting. They are merely the technology for resolving the interstate dispute."[25] This gave citizens little moral responsibility for the war, even when they were forced to participate, and helped to establish distinct norms of war existing apart from the norms of domestic life. The constant international debate over recent and ongoing wars is evidence that citizens do not feel this sense of detachment. Many are personally invested in conflicts, even participating in demonstrations and organizations that pressure states to avoid fighting. There is widespread suspicion that uncooperative civilians could interfere with wars, even bringing them to a premature end. Citizens must therefore be given reasons to support military operations. Sol-

25. Christopher Kutz, *On War and Democracy* (Princeton: Princeton University Press, 2016), 45.

diers must likewise be incentivized to participate in all-volunteer militaries. Failure to do so leaves states with inadequate manpower to prosecute wars, as happened when the United States was embroiled in Iraq and Afghanistan under the Bush administration.[26]

Because it is minimalistic and highly adaptable, degradation is marketable to civilian audiences and potential recruits. Wars with loosely defined and relatively limited goals can appeal to liberals who espouse humanitarian concerns as well as conservatives who want to reaffirm national prestige. Such wars can be fit into divergent narratives to appeal to the individual sensibilities of a range of different viewers. War must be transformed from a collective endeavor undertaken for disinterested reasons of state into an activity that can be molded to fit the assemblage of domestic and international constituencies that must be satisfied. Policymakers enact the logic of individuation to make war all things to all people. This cannot realistically satisfy everyone, but it certainly comes closer to that goal than would be possible if states adopted more heavy-handed conventional operations.

Renouncing Annihilation

Throughout the nineteenth and twentieth centuries the direction of technological and organizational changes was toward bigger and more powerful weapons and military formations. The driving motive behind innovation was to increase the destructive power that could be mobilized against enemies. The result was nuclear weapons and massive conventional militaries that were expensive to maintain and nearly impossible to use because of their immense destructive power. During the Cold War, any major deployment risked causing a global conflict and perhaps even global annihilation. Conflicts during that period were therefore marked by a turn toward more restricted wars aimed at achieving limited results. The United States avoided provoking the Soviet Union during its war in Vietnam. The Soviet Union avoided provoking the United States during its war in Afghanistan. International governing institutions and international agreements likewise formed to restrict the powerful weapons that resulted from the twentieth century's conventional wars, most notably nonproliferation agreements and conventions prohibiting certain weapons. Nevertheless, the threat of global war remained real and major innovations were heavily informed by

26. Lawrence J. Korb and Sean E. Duggan. "An All-Volunteer Army? Recruitment and Its Problems," *PS: Political Science and Politics* 40, no. 3 (2007): 467–71.

the dictates of conventional warfare. That is to say, innovation did not track well with changes in the missions armed forces were tasked with. This handicapped armed forces fighting insurgencies without adequate tools or training.

Degradation reflects an effort to learn from Cold War failures by reimagining the role of armed forces in a post–Cold War world. The restrictions on the use of force now come not only from the threat of retaliation but from domestic and international public opinion. Armed forces themselves are renouncing many of the technological and organizational remnants of conventional warfare. They are remaking themselves as small, lighter, agile, and adaptable forces. Most significantly of all, they are showing an interest in what Koplow labels "usable weapons," weapons that can be used to more easily launch strikes of a relatively low intensity compared to the conflicts states fought in the twentieth century. The goal of innovation is no longer to maximize destruction but to maximize effectiveness. And effectiveness depends on usability across a range of different conflict settings with diverse political objectives. This is a trend I discussed with respect to targeted killing, but here it is important to note that the same trend toward usability persists beyond operations directed against enemy leaders. Attacks on material infrastructure and even enemy units in the field are likewise marked by the goal of limiting destruction. Restraint and selectivity are now preferable even in interstate conflicts or operations against insurgents who are capable of fighting in pitched battles.

The pursuit of usable weapons for degrading operations has four major effects on militaries. First, as with other trends I have discussed, it increases the reliance on technology and the importance of research and development while also marginalizing the human participants in war. Fighting becomes a matter for technicians who manage the war from miles away. Even the pilots who are in the most direct contact with enemies have virtually no chance of being injured or killed and live fairly comfortable lives on bases stationed away from the fighting. This constitutes yet another simultaneous attack on the roles performed by military personnel and the culture of war. However, the threat here is slightly different from the one I have described in previous chapters. There is a tendency for members of the military to see counterinsurgency as something other than war and to view new technologies as aberrations. Targeted killings may likewise seem like something other than war, or perhaps a component of military operations, especially when these attacks are only undertaken by a small segment of military personnel. No such illusions can exist when it comes to degradation, as this style of fighting is not only introducing new facets of war but also directly contradicting the conventional war orientation. The persistent resort to degradation in interstate conflicts and conflicts with strong insurgencies signals that even operations that have previously fallen under the umbrella of conventional warfare are being reconfigured in ways that sharply conflict with the traditional military ethos. Moreover, conventional combat arms units are being remade with lighter, more agile weapons that fit with this style of fighting.[27]

Second, interest in degradation provides greater weight to the shift away from the kind of industrialized warfare that existed during the nineteenth and twentieth centuries. Many interstate wars closely resemble targeted killing and counterinsurgency operations and make use of the same technological and organizational capacities. The convergence of techniques in assassination and war provides incentives for continued investment in technologies and organizational structures that are amenable to both. This is particularly true when budgetary constraints compel maximum efficiency. States that pursue these as overlapping goals will gain a vested interest in continuing to model their operations along these lines, which will enact a path dependency to ensure that the innovations I discuss in this book will be prioritized well into the future.

Third, operations are shifting away from the combined arms doctrines that prevailed during the twentieth century. Combined arms warfare aspires to integrate all types of military forces into one seamless collective effort that takes the best advantage of relative strengths. Infantry, armor, and artillery work as a coordinated machine on land, while air and naval assets devastate enemies from a distance. Degradation, by contrast, tends to privilege one or two types of force apart from others, often with ground units being marginalized (except when it comes to special operations forces and contractors). In this sense, degradation stands in opposition to counterinsurgency. During counterinsurgency operations, naval and air forces struggle to carve out a role for themselves. When degradation is employed, it is ground forces that must struggle for orientation and to assert their own relevance. Counterinsurgency prioritizes long-term engagement in conflict areas and careful coordination with local fighters, while degradation tends to look more like a quick fix requiring minimal investment. Pursuit of two very different models of conflict pulls militaries in two different directions. One risk here is that rifts will emerge that limit capacities for joint operations. Another risk is that national security priorities will swing wildly back and forth between opposing mission types without an overriding sense of continuity or strategic

27. For example, the US Army's decision to replace heavy armored vehicles with Stryker Combat Vehicles.

vision. As ground forces adapt to counterinsurgency and attempt to find a place for themselves in degrading operations, they may render themselves unable to contend with threats posed by powerful rival states.

Finally, alongside counterinsurgency operations, degradation indicates that the world's most powerful militaries are shifting more to an education and advisement role. Rather than participating in pitched battles on the ground, soldiers train indigenous forces to do the fighting. They may advise operations and help to manage fire support assets, but they must avoid direct engagement. Even air and naval commitments may be reduced when efforts are made to displace security burdens onto coalition partners. This exacerbates the challenge to traditional military culture by further limiting opportunities for soldiers to distinguish themselves in combat. When seen alongside the overlapping challenges I have considered throughout the book, it is doubtful that militaries will be able to sustain existing conceptions of the warrior ethos and warrior values.

Conclusion

Degradation is primarily a strategy for advanced conventional armed forces like those of the United States and Israel. It is a strategy of the strong, pursued when those with power must impose limits on their own destructive force in the interest of being able to use military coercion without inviting total warfare. It is likewise closely associated with coalition conflicts in which it provides a minimalist strategic and political vision that can unify states with divergent interests. It assumes asymmetry between belligerents and the ability to launch precise strikes, which is why this new approach to war has become popular among aspiring global and regional hegemons (and their partners) since the end of the Cold War. It has been most clearly demonstrated by Presidents Clinton and Obama, which suggests an ideological appeal from the left, a desire to affect humanitarian change or to eliminate threats to national security without a high degree of militarization. It is likewise a style of fighting that reveals the influence of international public opinion, normative pressure, and the priority of political over military objectives.

The reliance on degradation fuels speculation that the United States and other Western democracies are casualty-averse and incapable of sustaining major ground campaigns. This sense that the United States would fall apart when fighting a more determined adversary was part of al Qaeda's rationale for attacking the United States and has been blamed for reducing US deterrence power. And yet degradation comes with greater freedom to launch military interventions. When attacks do not appear as acts of war, they can be launched more liberally and can be more easily legitimized. Moreover, when attacks are designed to achieve limited objectives, they can be reconciled with the conflicting goals of other interested parties, further lowering the barriers against fighting. Along with counterinsurgency and targeted killing, the preoccupation with degrading enemies points to a future of persistent restricted violence that ostensibly falls outside the boundaries of war. This will undermine the distinction between war and peace, as well as the geographical division of international and domestic spaces. Lower overall levels of violence and reduced wartime casualties will likely follow from this reconfiguration of war. Nevertheless, this practice is so heavily invested in methods of fighting that conflict with Westphalian conceptions of sovereignty and the international system that strong states are at risk of compromising their status over the long term.

Conclusion

Predictions about what future wars will look like are almost always wrong. Wars are simply too complex to be forecast accurately. I have stayed away from making claims about what will happen in future wars or creating a typology of future wars, and instead sought to show what tools state militaries will probably have when they enter those wars. Although the character of future wars is uncertain, the innovations I discussed are apt to continue on their present courses because states have demonstrated compelling interests in each. For example, it is virtually guaranteed that states will always find it advantageous to shield their personnel from danger by relying on drones as proxies. I argue that while it is impossible to predict exactly how the next war will be fought, it is possible to predict how militaries will be armed on the first day of that conflict and what direction institutional inertia will push them in based on current paths of innovation. Regardless of whether future wars are conventional or unconventional, symmetric or asymmetric, the participants will be shaped by the technological, organizational, and strategic innovations I have considered. Any future changes will have to build on this starting point and satisfy the same political demands. Subsequent developments will not constitute a clean break with the past but instead a change of direction in ongoing developmental processes.

The innovations I have discussed demonstrate that war is an increasingly ambiguous concept and that the conventional war vs. peace binary at the heart of the Westphalian system is untenable. Although the innovations can be used in many different conflict types, they derive special theoretical significance from how they facilitate violence that is difficult to classify within existing frameworks. It is often in the space between war and peace that states have the best hopes of achieving their objectives without triggering an escalation of hostilities or backlash from constituents. However, when states operate in this way, they undercut the Westphalian system and their own sovereign authority. As the concepts of war and peace attenuate, so too does an understanding of politics that treated conventional wars as a standard method of resolving disagreements between states and domestic spaces as falling under the sole authority of a particular sovereign. This is an ironic outcome in two ways. First, the extreme danger of escalating hostilities, which pushes states to use force beyond conventional wars, is a product of states' own unrestrained innovations creating arsenals of such power that they are no longer usable. Second, states are pursuing the kinds of innovations that extend conflict beyond conventional war in ways that secure short-term advantages at the expense of building a more fundamental long-term challenge to the Westphalian system. This makes the threat to the Westphalian system as much a product of state actions as it is the product of security hazards created by nonstate actors.

War is no longer a normatively and politically exceptional state of affairs. It is a background condition of life that persists through a range of conflict types, from conventional fighting that consistently happens without declarations of war, to low-intensity operations that resemble law enforcement, to law enforcement operations that make use of military hardware. On any particular day, there may be cyberattacks between rival states, small battles between their respective private contractors, the use of nonlethal weapons to maintain order in occupied areas, targeted attacks to eliminate key enemy personnel, and population-centric campaigns to attract weakly aligned civilians and enemy fighters. In the background are ongoing debates about whether the kinds of innovations I discuss can be treated as instruments of war, which predictably result in deadlock because of the inherent challenges in imposing a clear form on something that is formless, often by design. I have evaluated the innovations with help from seven interlocking themes as a way of highlighting the key theoretical implications without relying on strict typologies that would merely reproduce the conceptual barriers. At a time when typologies are unhelpful, it is possible to gain some degree of clarity by investigating war with themes that establish family resemblances between the various activities that constitute war.

A preoccupation with managing risks shows that states are aware of their precarious status at a time when the future of the international system is in doubt, Westphalian customs of political action are not followed, and myriad nonstate actors create novel security challenges. By investing in technologies, organizational structures, and strategies that can be used across a range of potential future conflict scenarios, states cultivate resilience. They seek to be prepared for all possible eventualities, not merely for conventional war. Militaries transform from specialists in overwhelming violence to eclectic forces capable of adapting to any mission. Moreover, risk calculations extend the scope of security decisions beyond conventional goals such as preventing attacks or defeating enemies. States aspire to nothing less than total control over the consequences of use-of-force decisions for anyone who may be affected by them. This is evident from how innovations are designed to give control over which specific adversaries are killed and whether they are killed at all.

Physical mediation has reached a point where conflicts can take place at virtually any distance. Drone operators and cyberwarriors can fight anywhere in the world, and without leaving home countries that do not look like active war zones. The quest for new forms of mediation have gone so far as to change the physical substance of war at the most fundamental level, expanding violence from blunt instruments and projectiles to gases, energy beams, sound waves, and more. Such dramatic changes in the form of war cannot help but impact how war is conceptualized. Institutional mediation has similar effects when it comes to special operations forces, private contractors, and the third-party forces that are key to population-centric counterinsurgencies. Because of their secrecy and distance from decision makers, these combatants can be sent on missions far beyond conventional war zones, and often with limited visibility. Physical and institutional mediation jointly make it more difficult to attribute responsibility for security decisions. They allow states to make war with a degree of deniability. While this deniability is not always plausible, and at times stretches credulity, it tends to at least obfuscate decisions enough that risks associated with public opinion can be minimized.

One of the foremost risks that states must confront is that their security decisions will have high public opinion costs. New communications technologies, including weapons whose physical mediation creates images for mass consumption, give domestic and international audiences an intimate look at ongoing conflicts. New communications technologies likewise make it possible to quickly build and circulate alternative narratives to impart meaning to those conflicts. Declining domestic support can pull resources away from the war effort and compel policymakers to shift their strategies. Declining international support can weaken alliances and provoke reprisals. States must correspondingly extend their wars into a communicative domain and wrest control of the narrative by managing audience perceptions. Physical and institutional mediation are among the central tools in this effort, as they provide layers of insulation from potentially unpopular decisions.

Although war continues to take place in physical spaces, the political meaning of those spaces has changed. Lack of respect for geographical boundaries has altered the territorial dimension of war, and with it the territorial dimension of politics. With battle spaces that transcend national borders, combatants fighting from extreme distances, and no clear line between territories of war and peace, the physical reference points that once gave the Westphalian system a sense of permanence are disappearing. Even outwardly territorially conservative campaigns that aim to degrade enemies without remaking borders are conducted in ways that subvert the Westphalian conception of sovereignty, speeding its demise. The most significant development here is that sovereign authority within state borders is no longer stable; it is contestable. Weak states are frequently transformed into complex battle spaces where various outside actors attempt to impose their will. Even states with relatively strong conventional militaries have limited power to maintain authority within their territories when foreign adversaries carry out attacks with mediated weapons and organizations. The kinds of territorial intrusions that were once cause for conventional war have become a matter of routine politics.

Military service has gone from a type of collective service to the state to a niche occupation, giving militaries of the twenty-first century a much different composition than those of the nineteenth and twentieth centuries. Moreover, the boundaries of the military profession are disappearing and will become more difficult to identify as private actors, civilian government officials, and technical specialists play a larger role in operations. The resort to private assistance is a long-term hindrance that states have imposed on themselves in pursuit of short-term risk-management gains. Increased involvement of civilians in military operations is likewise self-imposed: a byproduct of wars aimed at achieving more lasting and fundamental political objectives, such as regime change, nation building, and eliminating the root causes of terrorism. Challenges are also emerging from within the ranks as the number of soldiers participating in direct combat decreases and the number of technical specialists involved in physically mediated violence increases. With the military profession's boundaries disappearing, and new internal divisions emerging, militaries of the future are apt to be more heterogeneous and open-ended than those of the Westphalian era.

To a degree unprecedented since pre-Westphalian conflicts, combatants participate in wars with their individuality intact. Armed forces are less inclined to subject their members to the kind of rigid, authoritarian control that prevailed during the height of the Westphalian system. When it comes to the highly specialized roles I have discussed throughout the book, militaries lack the ability to impose such control. Special operations forces and population-centric counterinsurgency presuppose devolution of authority to individuals. Similarly, roles that demand a high degree of technical skill possessed by relatively few people are rarely amenable to strict control. Individualism is projected onto enemies as well. They are disaggregated into more and less dedicated fighters, those who must be fought versus those who can be co-opted. The former become the targets of selective violence in targeted killing and degradation campaigns. The latter become audience members whose perceptions must be carefully managed.

States are losing the monopoly on legitimate force that was the hallmark of the Westphalian state, often through their own decisions to purchase private assistance, empower technical specialists, and redefine military missions in ways that politicize and moralize the participants. They are therefore losing the grounds for claiming that state violence has a higher status—morally or politically—than violence promulgated by other types of actors. Norms coming from the just war tradition, systems of military ethics, and international law, which have been molded to fit wars within the Westphalian system, are rife with contradictions in the post-Westphalian era. When violence is difficult to classify as war or something else, it is difficult to say whether norms of war or those associated with law enforcement should pertain. And when wars are waged in ways that do not conform to the conventional model, existing norms provide imperfect guidance.

At best, changes in the composition of armed forces and the conduct of war bring uncertainty, which calls for a re-evaluation of what war is and how it should be situated in relation to other political activities. At worst, these changes could substantively undermine security and cause more human suffering. Throughout the book I have sought to show that the innovations are still at a point where the implications can be shaped by policy decisions. Many of the innovations even suggest a positive developmental trajectory. For example, population-centric operations could promote greater care for civilians, and nonlethal weapons could reduce the number of civilians killed. On the other hand, all the innovations have the potential for misuse. The expansion of conflict beyond neatly confined boundaries is itself deeply threatening because it forecloses the possibility of peaceful politics beyond war. Theorizing these innovations is a first step toward understanding the range of positive and negative implications, which I hope lays a more promising foundation for efforts to manage their use.

Bibliography

Abrahms, Max. "The Political Effectiveness of Terrorism Revisited." *Comparative Political Studies* 45, no. 3 (2012): 366–93.

Agwu, Fred Aja. *Armed Drones and Globalization in the Asymmetric War on Terror: Challenges for the Law of Armed Conflict and Global Political Economy*. London: Routledge, 2017.

Alexander John B. *Future War: Nonlethal Weapons in Twenty-First-Century Warfare*. New York: St. Martin's Press, 1999.

Altmann, Jürgen. "Acoustic Weapons: A Prospective Assessment." *Science & Global Security* 9 (2001): 165–234.

Aylwin-Foster, Nigel. "Changing the Army for Counterinsurgency Operations." *Military Review* (November–December 2005).

Arkin, Ronald C. *Governing Lethal Behavior in Autonomous Robots*. Boca Raton, FL: Taylor & Francis, 2009.

Arnold, Thomas. *The Renaissance at War*. London: Cassell, 2001.

Ashworth, Tony. *Trench Warfare 1914–1918: The Live and Let Live System*. London: Macmillan, 1980.

Asprey, Robert. *War in the Shadows: The Guerrilla in History*. iUniverse, 2002.

Atkeson, Lonna Rae, and Cherie D. Maestas. *Catastrophic Politics: How Extraordinary Events Redefine Perceptions of Government*. Cambridge: Cambridge University Press, 2012.

Axelrod, Alan. *Mercenaries: A Guide to Private Armies and Private Military Companies*. Washington, DC: CQ Press, 2014.

Bacevich, Andrew J. *America's War for the Greater Middle East: A Military History*. New York: Random House, 2016.

Baer, Robert B. *The Perfect Kill: 21 Laws for Assassins*. New York: Plume, 2014.

Baker, Anni. *Life in the U.S. Armed Forces: (Not) Just Another Job*. London: Praeger, 2008.

Ballard, John R. *Fighting for Fallujah: A New Dawn for Iraq*. Westport: CT: Praeger, 2006.

Baudrillard, Jean. *The Gulf War Did Not Take Place*. Bloomington: Indiana University Press, 1995.

Beaumont, Roger. *Military Elites*. New York: Bobbs-Merrill, 1974.

Beck, Ulrich. *World Risk Society*. Cambridge: Polity, 1999.

Beevor, Antony. *Ardennes 1944: The Battle of the Bulge*. New York: Penguin, 2016.

Bell, Colleen "Civilianising Warfare: Ways of War and Peace in Modern Counterinsurgency." *Journal of International Relations and Development* 14, no. 3 (2011): 309–32.

Bencsáth, Boldizsár, Gábor Pék, Levente Buttyán, and Mark Felegyhazi. "The Cousins of Stuxnet: Duqu, Flame, and Gauss." *Future Internet* 4, no. 4 (2012).

Benjamin, Medea. *Drone Warfare: Killing by Remote Control*. New York: OR Books, 2012.

Bennett, Huw. *Fighting the Mau Mau: The British Army and Counter-Insurgency in the Kenya Emergency*. Cambridge: Cambridge University Press, 2013.

Bergman, Ronen. *Rise and Kill First: The Secret History of Israel's Targeted Assassinations*. New York: Random House, 2018.

Black, Jeremy. *Warfare in the Eighteenth Century*. London: Cassell, 2002.

Bond, Brian. *The Pursuit of Victory*. New York: Oxford University Press, 1998.

Boot, Max. *Invisible Armies*. New York: Liveright, 2013.

Boudreau, Tyler E. *Packing Inferno: The Unmaking of a Marine*. Port Townsend, WA: Feral House, 2008.

Bourke, Joanna. *An Intimate History of Killing: Face-to-Face Killing in Twentieth-Century Warfare*. London: Granta Books, 1999.

Bowden, Mark. *Black Hawk Down: A Story of Modern War*. New York: Atlantic Monthly Press, 1999.

Bowden, Mark. *The Finish: The Killing of Osama bin Laden*. Washington, DC: Atlantic Monthly Press, 2012.

Boys, James D. *Clinton's Grand Strategy: US Foreign Policy in a Post-Cold War World*. London: Bloomsbury, 2015.

Branch, Daniel. "Footprints in the Sand: British Colonial Counterinsurgency and the War in Iraq." *Politics & Society* 38, no. 1 (2010): 15–34.

Bregman, Ahron. *Israel's Wars: A History since 1947*. London: Routledge, 2016.

British Army Field Manual. Vol. 1, part 10, *Countering Insurgency* (October 2009).

Brooks, Doug, and Matan Chorev. "Ruthless Humanitarianism: Why Marginalizing Private Peacekeeping Kills People." In *Private Military and Security Companies: Ethics, Policies and Civil-Military Relations*, edited by A. Alexandra, D-P. Baker, and M. Caparini, 116–30. New York: Routledge, 2008.

Burk, James. "Public Support for Peacekeeping in Lebanon and Somalia: Assessing the Casualties Hypothesis." *Political Science Quarterly* 114, no. 1 (1999): 53–78.

Carafano, James Jay. *Private Sector, Public Wars: Contractors in Combat - Afghanistan, Iraq, and Future Conflicts*. Santa Barbara, CA: Praeger, 2008.

Carmola, Kateri. *Private Security Contractors and New Wars: Risk, Law, and Ethics*. London: Routledge, 2010.

Carruthers, Susan L. *The Media at War*. New York: Palgrave Macmillan, 2011.

Catignani, Sergio. *Israeli Counter-Insurgency and the Intifadas: Dilemmas of a Conventional Army*. New York: Routledge, 2008.

Center for Army Lessons Learned. *Traffic Control Point Operations (GTA 90-01-005)* (2010).

Challans, Timothy L. *Awakening Warrior: Revolution in the Ethics of Warfare*. Albany: State University of New York Press, 2007.

Chamayou, Grégorie. *A Theory of the Drone*. New York: The New Press, 2015.
Chin, Warren. "Examining the Application of British Counterinsurgency Doctrine by the American Army in Iraq." *Small Wars & Insurgencies* 18, no. 1 (2007): 1–26.
Chivvis, Christopher S. *Toppling Qaddafi: Libya and the Limits of Liberal Intervention*. Cambridge: Cambridge University Press, 2013.
Clausewitz, Carl von. *On War*. Oxford: Oxford University Press, 2007.
Clemis, Martin G. "Crafting Non-Kinetic Warfare: The Academic-Military Nexus in US Counterinsurgency Doctrine." *Small Wars & Insurgencies* 20, no. 1 (2009): 160–84.
Coady, C. A. J. "Mercenary Morality." In *International Law and Armed Conflict*, edited by A. G. D. Bradney, 55–69. Stuttgart: Franz Steiner Verlag, 1992.
Cohen, Eliot A. *Commandos and Politicians: Elite Military Units in Modern Democracies*. Cambridge, MA: Harvard Center for International Affairs, 1978.
Coker, Christopher. *Waging War Without Warriors?* Boulder: Lynne Rienner, 2002.
Coll, Steve. *Directorate S: The CIA and America's Secret Wars in Afghanistan and Pakistan, 2001–2016*. New York: Penguin, 2018.
Connelly, James, Don Carrick, and Paul Robinson, eds. *Ethics Education for Irregular Warfare*. Burlington, VT: Ashgate, 2009.
Cook, James. "'Cyberation' and Just War Doctrine: A Response to Randall Dipert." *Journal of Military Ethics* 9, no. 4 (2010): 411–23.
Cook, James. *The Moral Warrior: Ethics and Service in the U.S. Military*. Albany: State University of New York Press, 2004.
Cordesman, Anthony H. *The Iraq War: Strategy, Tactics, and Military Lessons*. Westport, CT: Greenwood, 2003.
Couch, Dick. *Down Range: Navy SEALs in the War on Terrorism*. New York: Three Rivers Press, 2006.
Couch, Dick. *A Tactical Ethic: Moral Conduct in the Insurgent Battlespace*. Annapolis, MD: Naval Institute Press, 2011.
Davenport, Manuel M. "Professionals or Hired Guns? Loyalties Are the Difference" In *Military Ethics: Reflections on Principles*, ed. Malham M. Wakin. Washington, DC: National Defense University Press, 1987.
Davis, Richard G. *Decisive Force: Strategic Bombing in the Gulf War*. Air Force History and Museums Program, 1996.
David, Steven R. "Israel's Policy of Targeted Killing." *Ethics & International Affairs* 17, no. 1 (2003): 111–26.
Deptula, David A. *Effects-Based Operations: Change in the Nature of Warfare*. Arlington, VA: Aerospace Education Foundation, 2001.
Der Derian, James. *Virtuous War: Mapping the Military-Industrial-Media-Entertainment Network*. New York: Routledge, 2009.
DeVries, Kelly, and Robert Douglas Smith. *Medieval Military Technology*. Toronto: University of Toronto Press, 2012.
Downes, Cathy. "To Be Or Not to Be a Profession: The Military Case." *Defence Analysis* 1, no. 3 (1985): 147–71.
Dyer, Gwynne. *War: The Lethal Custom*. New York: Carroll & Graf, 1984.
Ehrenreich, Barbara. *Blood Rites: Origins and History of the Passions of War*. New York: Henry Holt, 1997.
Ehrhard, Thomas P. *Air Force UAVs: The Secret History*. Arlington, VA: Mitchell Institute for Airpower Studies, 2010.

Ellis, John. *Brute Force: Allied Strategy and Tactics in the Second World War*. London: Andre Deutsch, 1990.

Eun, Yong-Soo, and Judith Sita Aßmann. "Cyberwar: Taking Stock of Security and Warfare in the Digital Age." *International Studies Perspectives* 17, no. 3 (2016): 343–60.

Fahlstrom, Paul, and Thomas Gleason. *Introduction to UAV Systems*. Hoboken, NJ: Wiley, 2012.

Fainaru, Steve. *Big Boy Rules: America's Mercenaries Fighting in Iraq*. Philadelphia: Da Capo Press, 2008.

Farwell, James P., and Rafal Rohozinski. "Stuxnet and the Future of Cyber War." *Survival: Global Politics and Strategy* 53, no. 1 (2011): 23–40.

Feigenbaum, Anna. *Tear Gas: From the Battlefields of WWI to the Streets of Today*. New York: Verso Books, 2017.

Finer, S. E. *The Man on Horseback: The Role of the Military in Politics*. Boulder: Westview Press, 1988.

FM 3-24/MCWP 3-33.5 Counterinsurgency. Washington, DC: Headquarters Department of the Army, 2006.

Foot, Michael, and Richard Daniell. *SOE: An Outline History of the Special Operations Executive 1940–46*. New York: Random House, 2011.

Fotion, Nicholas, and Gerard Elfstrom. *Military Ethics: Guidelines for Peace and War*. New York: Routledge, 1986.

France, John, ed. "Mercenaries and Paid Men: The Mercenary Identity in the Middle Ages." Proceedings of a Conference Held at University of Wales, Swansea, July 7–9, 2005. Vol. 47. Leiden: Brill, 2008.

French, Shannon E. *The Code of the Warrior: Exploring Warrior Values Past and Present*. New York: Rowman & Littlefield, 2003.

Gartzke, Erik, and Jon R. Lindsay. "Weaving Tangled Webs: Offense, Defense, and Deception in Cyberspace." *Security Studies* 24, no. 2 (2015): 316–48.

Gelpi, Christopher, Peter D. Feaver, and Jason Reifler. "Success Matters: Casualty Sensitivity and the War in Iraq." *International Security* 30, no. 3 (2006): 7–46.

Gelpi, Christopher, Peter D. Feaver, and Jason Reifler. *Paying the Human Costs of War: American Public Opinion and Casualties in Military Conflicts*. Princeton: Princeton University Press, 2009.

Gentile, Gian P. "A Strategy of Tactics: Population-centric COIN and the Army." *Parameters* 39, no. 3 (2009): 5–17.

Giddens, Anthony. *Modernity and Self-Identity: Self and Society in the Late Modern Age*. Cambridge: Polity Press, 1991.

Gilmore, Jonathan. "A Kinder, Gentler Counter-Terrorism: Counterinsurgency, Human Security and the War on Terror." *Security Dialogue* 42, no. 21 (2011): 21–37.

Gonzales, Roberto J. *Militarizing Culture: Essays on the Warfare State*. Walnut Creek, CA: Left Coast Press, 2010.

Graham, Stephen. *Cities Under Siege: The New Military Urbanism*. New York: Verso, 2010.

Graham, Stephen. "When Life Itself Is War: On the Urbanization of Military and Security Doctrine." *International Journal of Urban and Regional Research* 36, no. 1 (2012): 136–55.

Gray, Glenn J. *The Warriors: Reflections on Men in Battle*. New York: Harper & Row, 1967.

Grayling, A. C. *Among the Dead Cities: The History and Moral Legacy of the WWII Bombing of Civilians in Germany and Japan*. New York: Walker & Company, 2006.

Greenberg, Andy. *Sandworm*. New York: Anchor, 2019.

Gregory, Derek. "From a View to a Kill: Drones and Late Modern War." *Theory, Culture, & Society* 28, no. 7–8 (2011): 188–215.

Gross, Michael L. "Assassination and Targeted Killing: Law Enforcement, Execution or Self-Defence?" *Journal of Applied Philosophy* 23, no. 3 (2006): 323–35.

Gross, Michael L. *Moral Dilemmas of Modern War: Torture, Assassination, and Blackmail in an Age of Asymmetric Conflict*. New York: Cambridge University Press, 2009.

Halberstam, David. *War in a Time of Peace: Bush, Clinton, and the Generals*. New York: Simon & Schuster, 2001.

Hanson, Victor Davis. *The Western Way of War: Infantry Battle in Classical Greece*. New York: Random House, 2013.

Hassig, Ross. *Aztec Warfare: Imperial Expansion and Political Control*. Norman: University of Oklahoma Press, 1988.

Hazell, Paul J. *Armour: Materials, Theory, and Design*. Boca Raton, FL: CRC Press, 2016.

Hedges, Chris. *War Is a Force That Gives Us Meaning*. New York: PublicAffairs, 2002.

Hegel, Georg. *The Science of Logic*. Cambridge: Cambridge University Press, 2010.

Heinecken, Lindy. "Outsourcing Public Security: The Unforeseen Consequences for the Military Profession," *Armed Forces & Society* 40, no. 4 (2014): 625–46.

Hess, Earl J. *Civil War Logistics* (Baton Rouge: Louisiana State University Press, 2017).

Holmqvist, Caroline. "Undoing War: War Ontologies and the Materiality of Drone Warfare." *Millennium* 41, no. 3 (2013): 535–52.

Holsti, Kalevi. *The State, War, and the State of War*. New York: Cambridge University Press, 1996.

Hood, Christopher. *The Blame Game: Spin, Bureaucracy, and Self-Preservation in Government*. Princeton: Princeton University Press, 2010.

Horowitz, Michael C., Sarah E. Kreps, and Matthew Fuhrmann. "Separating Fact from Fiction in the Debate over Drone Proliferation." *International Security* 41, no. 2 (2016).

Hoskins, Andrew, and Ben O'Loughlin. *War and Media*. Cambridge: Polity Press, 2010.

Huntington, Samuel P. *The Soldier and the State: The Theory and Politics of Civil-Military Relations*. Cambridge, MA: Harvard University Press, 1959.

Hurley, Matthew M. "The Bekaa Valley Air Battle." *Airpower Journal* (Winter 1989).

Hyde, Charles K. *Casualty Aversion: Implications for Policy Makers and Senior Military Officers*. Newport, RI: Naval War College, 2000.

Ignatieff, Michael. *Virtual War: Kosovo and Beyond*. New York: Picador, 2001.

Iyengar, Shanto. *Is Anyone Responsible? How Television Frames Political Issues*. Chicago: University of Chicago Press, 1994.

Janowitz, Morris. *The Professional Soldier: A Social and Political Portrait*. New York: The Free Press, 1960.

Jentleson, Bruce W. "The Pretty Prudent Public: Post Post-Vietnam American Opinion on the Use of Military Force." *International Studies Quarterly* 36, no. 1 (1992): 49–73.

Jentleson, Bruce W., and Rebecca L. Britton. "Still Pretty Prudent: Post–Cold War American Public Opinion on the Use of Military Force." *Journal of Conflict Resolution* 42, no. 4 (1998): 395–417.

Joes, Anthony James. *Urban Guerrilla Warfare*. Lexington: University Press of Kentucky, 2007.

Johnson, James Turner. *Just War Tradition and the Restraint of War: A Moral and Historical Inquiry*. Princeton: Princeton University Press, 1981.

Joint Chiefs of Staff. *Joint Urban Operations, 3-06*. Washington, DC: Joint Chiefs of Staff. 2013.

Jordan, Jenna. "When Heads Roll: Assessing the Effectiveness of Leadership Decapitation." *Security Studies* 18, no. 4 (2009): 719–55.

Kaldor, Mary. *Old and New Wars: Organized Violence in a Global Era*. Stanford: Stanford University Press, 1998.

Kalyvas, Stathis N. "The Ontology of 'Political Violence': Action and Identity in Civil Wars." *Perspectives on Politics* 1, no. 3 (2003): 475–94.

Kaplan, Fred. *Dark Territory: The Secret History of Cyber War*. New York: Simon & Schuster, 2016.

Katz, Yaakov, and Amir Bohbot. *The Weapon Wizards: How Israel Became a High-Tech Military Superpower*. New York: St. Martin's Press, 2017.

Kaufmann, Walter. *Nietzsche: Philosopher, Psychologist, Antichrist*. Princeton: Princeton University Press, 1974.

Kaurin, Pauline. "With Fear and Trembling: An Ethical Framework for Nonlethal Weapons." *Journal of Military Ethics* 9, no. 1 (2010): 100–114.

Keegan, John. *A History of Warfare*. New York: Vintage, 1994.

Keen, Maurice. *Chivalry*. New Haven: Yale University Press, 2005.

Kennedy, David. *Of War and Law*. Princeton: Princeton University Press, 2006.

Kilcullen, David. *The Accidental Guerrilla: Fighting Small Wars in the Midst of a Big One*. New York: Oxford University Press, 2009.

Kilcullen, David. *Blood Year: Islamic State and the Failures of the War on Terror*. Oxford: Oxford University Press, 2016.

Kilcullen, David. *Out of the Mountains: The Coming Age of the Urban Guerrilla*. New York: Oxford University Press, 2013.

Kinsella, Helen M. *The Image Before the Weapon: A Critical History of the Distinction Between Combatant and Civilian*. Ithaca: Cornell University Press, 2011.

Kittrie, Orde F. *Lawfare: Law as a Weapon of War*. Oxford: Oxford University Press, 2016.

Koplow, David A. *Death by Moderation: The U.S. Military's Quest for Useable Weapons*. New York: Cambridge University Press, 2010.

Koplow, David A. *Nonlethal Weapons: The Law and Policy of Revolutionary Technologies for the Military and Law Enforcement*. Cambridge: Cambridge University Press, 2006.

Korb, Lawrence J., and Sean E. Duggan. "An All-Volunteer Army? Recruitment and Its Problems." *PS: Political Science and Politics* 40, no. 3 (2007): 467–71.

Krahman, Elke. "The United States, PMSCs and the State Monopoly on Violence: Leading the Way Towards Norm Change." *Security Dialogue* 44, no. 1 (2013): 53–71.

Krulak, Charles C. "The Strategic Corporal: Leadership in the Three Block War." *Marines Magazine* 28 (1999). http://www.au.af.mil/au/awc/awcgate/usmc/strategic_corporal.htm

Kutz, Christopher. *On War and Democracy*. Princeton: Princeton University Press, 2016.

Kyle, Chris, Scott McEwen, and Jim DeFelice. *American Sniper: The Autobiography of the Most Lethal Sniper in US Military History*. New York: William Morrow, 2013.

Lanoszka, Alexander. "Russian Hybrid Warfare and Extended Deterrence in Eastern Europe." *International Affairs* 92, no. 1 (2016): 175–95.

Larson, Eric V. *Casualties and Consensus: The Historical Role of Casualties in Domestic Support for U.S. Military Operations*. Santa Monica, CA: RAND, 1996.

Ledwidge, Frank. *Losing Small Wars: British Military Failure in Iraq and Afghanistan*. New Haven: Yale University Press, 2011.

Lee, Graham. *Fighting Season*. London: Duckworth Overlook, 2012.

Lee, Wayne E. *Waging War: Conflict, Culture, and Innovation in World History*. New York: Oxford University Press, 2016.

Lucas, George R. *Military Ethics: What Everyone Needs to Know*. New York: Oxford University Press, 2016.

Lucas, James Sidney. *Storming Eagles: German Airborne Forces in World War Two*. London: Arms & Armour Press, 1988.

Luttrell, Marcus. *Lone Survivor: The Incredible True Story of Navy SEALs Under Siege*. London: Sphere, 2014.

Luttwak, Edward N. *Strategy: The Logic of War and Peace*. Cambridge, MA: Belknap Press, 2001.

Luttwak, Edward N. "Toward Post-Heroic Warfare." *Foreign Affairs* (May/June 1995).

Lynn, John A. *Battle: A History of Combat and Culture*. New York: Basic Books, 2003.

Machiavelli, Niccolò. *The Art of War*. New York: Barnes and Noble, 2004.

Maestas, Cherie D., et al. "Shifting the Blame: Federalism, Media, and Public Assignment of Blame Following Hurricane Katrina." *Publius: The Journal of Federalism* 38, no. 4 (2008): 609–32.

Mallett, Michael. *Mercenaries and their Masters: Warfare in Renaissance Italy*. Barnsley: Pen & Sword, 2009.

Mansoor, Peter R. *Surge: My Journey with General David Petraeus and the Remaking of the Iraq War*. New Haven: Yale University Press, 2013.

Martel, William C. *Victory in War: Foundations of Modern Strategy*. Cambridge: Cambridge University Press, 2011.

Martin, Matt J., and Charles W. Sasser. *Predator: The Remote-Control Air War Over Iraq and Afghanistan: A Pilot's Study*. Minneapolis: Zenith Press, 2010.

Masland, John Wesley, and Laurence I. Radway. *Soldiers and Scholars: Military Education and National Policy*. Princeton: Princeton University Press, 2015.

Maurer, Tim. *Cyber Mercenaries: The State, Hackers, and Power*. Cambridge: Cambridge University Press, 2018.

McCoy, Alfred. *A Question of Torture: CIA Interrogation, from the Cold War to the War on Terror*. New York: Metropolitan Books, 2007.

McFate, Montgomery. "Anthropology and Counterinsurgency: The Strange Story of Their Curious Relationship," *Military Review* 85, no. 2 (2005): 24–38.

McFate, Sean. *The Modern Mercenary: Private Armies and What They Mean for World Order*. Oxford: Oxford University Press, 2017.

McGrath, John J. *The Other End of the Spear: The Tooth-to-Tail Ratio (T3R) in Modern Military Operations*. Fort Leavenworth, KS: Combat Studies Institute Press, 2011.

McInnes, Colin. *Spectator Sport War: The West and Contemporary Conflict*. Boulder: Lynne Rienner, 2002.

McNab, Robert M., and Richard L. Scott. "Nonlethal Weapons and the Long Tail of Warfare." *Small Wars & Insurgencies* 20, no. 1 (2009): 141–59.

McRaven, William H. "Preparing Special Operations Forces for the Future." *Journal of International Security Affairs* 23 (Fall 2012).

McRaven, William H. *Spec Ops: Case Studies in Special Operations Warfare: Theory and Practice*. New York: Presidio Press, 2009.

Micheletti, Eric. *Special Forces: War Against Terrorism in Afghanistan, 2001–2003*. Paris: Histoire & Collections, 2003.

Mitnick, Kevin D., and William L. Simon. *The Art of Deception*. Indianapolis: Wiley, 2002.

Mitnick, Kevin D., and William L. Simon. *The Ghost in the Wires: My Adventures as the World's Most Wanted Hacker*. New York: Little, Brown, 2011.

Mockaitis, Thomas R. *The COIN Conundrum: The Future of Counterinsurgency and US Land Power*. Carlisle Barracks, PA: US Army War College–Strategic Studies Institute, 2016.

Moskos, Charles C. Jr. *The American Enlisted Man: The Rank and File in Today's Military*. New York: Russell Sage Foundation, 1970.

Mrazek, James E. *Airborne Combat: The Glider War/Fighting Gliders of WWII*. Mechanicsburg, PA: Stackpole Books, 2011.

Mueller, John. *Overblown: How Politicians and the Terrorism Industry Inflate National Security Threats, and Why We Believe Them*. New York: Simon & Schuster, 2009.

Mueller, John. *Policy and Opinion in the Gulf War*. Chicago: University of Chicago Press, 1994.

Mueller, John. *War, Presidents and Public Opinion*. New York: Wiley, 1973.

Münkler, Herfried. *The New Wars*. Malden, MA: Polity Press, 2005.

Mythen, Gabe, and Sandra Walklate. "Terrorism, Risk and International Security: The Perils of Asking 'What If?'" *Security Dialogue* 39, no. 2–3 (2008): 221–42.

Nagl, John. *Learning to Eat Soup with a Knife: Counterinsurgency Lessons from Malaya and Vietnam*. Chicago: University of Chicago Press, 2005.

Naylor, Sean. *Relentless Strike: The Secret History of Joint Special Operations Command*. New York: St. Martin's Griffin, 2016.

Neff, Stephen. *War and the Law of Nations: A General History*. Cambridge: Cambridge University Press, 2005.

Neville, Leigh. *Special Operations Forces in Afghanistan*. London: Bloomsbury, 2011.

Nye, Joseph S. *Soft Power: The Means to Success in World Politics*. New York: Public Affairs, 2004.

O'Donnell, Patrick K. *Operatives, Spies, and Saboteurs: The Unknown Story of the Men and Women of World War II's OSS*. New York: Simon & Schuster, 2014.

O'Donnell, Patrick. *We Were One: Shoulder to Shoulder with the Marines Who Took Fallujah*. New York: Da Capo Press, 2007.

Olson, Parmy. *We Are Anonymous: Inside the Hacker World of LulzSec, Anonymous, and the Global Cyber Insurgency*. New York: Little, Brown, 2012.

O'Neill, Robert. *The Operator: Firing the Shots that Killed Osama Bin Laden and My Years as a SEAL Team Warrior*. New York: Scribner, 2017.

Orend, Brian. "*Jus Post Bellum*: The Perspective of a Just-War Theorist." *Leiden Journal of International Law* 20, no. 3 (2007): 571–91.

Orend, Brian. *The Morality of War*. Orchard Park, NY: Broadview Press, 2006.

Orend, Brian. *War and International Justice: A Kantian Perspective.* Waterloo, ON: Wilfrid Laurier University Press, 2000.

Owen, Mark, and Kevin Maurer. *No Easy Day: The Firsthand Account of the Mission That Killed Osama Bin Laden.* New York: Penguin, 2014.

Parrott, David *The Business of War: Military Enterprise and Military Revolution in Early Modern Europe.* New York: Cambridge University Press, 2012.

Pegler, Martin. *Out of Nowhere: A History of the Military Sniper.* Oxford: Osprey, 2004.

Percy, Sarah. *Mercenaries: The History of a Norm in International Relations.* New York: Oxford University Press, 2007.

Petton, Robert Young. *Licensed to Kill: Privatizing the War on Terror.* New York: Three Rivers Press, 2006.

Phillips, David L. *Losing Iraq: Inside the Postwar Reconstruction Fiasco.* New York: Basic Books, 2009.

Philpott, William. *Attrition: Fighting the First World War.* London: Little, Brown, 2014.

Porch, Douglas. *Counterinsurgency: Exposing the Myths of the New Way of War.* Cambridge: Cambridge University Press, 2013.

Prince, Erik. *Civilian Warriors: The Inside Story of Blackwater and the Unsung Heroes of the War on Terror.* New York: Penguin, 2013.

Rabinovich, Abraham. *The Yom Kippur War.* New York: Schocken, 2004.

Rappert, Brian. *Nonlethal Weapons as Legitimizing Forces? Technology, Politics, and the Management of Conflict.* London: Frank Cass, 2003.

Rasmussen, Mikkel Vedby. *The Risk Society at War: Terror, Technology and Strategy in the Twenty-First Century.* New York: Cambridge University Press, 2006.

Robinson, Linda. *The Future of US Special Operations Forces.* New York: Council on Foreign Relations, 2013.

Robinson, Paul. "Ethics Training and Development in the Military." *Parameters* (2007).

Robinson, Paul, Nigel De Lee, Don Carrick, ed. *Ethics Education in the Military.* Hampshire: Ashgate, 2008.

Rodin, David. *War and Self-Defense.* New York: Oxford University Press, 2003.

Rosen, Stephen Peter. *Winning the Next War: Innovation and the Modern Military.* Ithaca: Cornell University Press, 1991.

Rothenberg, Gunther E. *The Art of Warfare in the Age of Napoleon.* Indianapolis: Indiana University Press, 1978.

Rottman, Gordon L. *Vietnam Airmobile Warfare Tactics.* London: Bloomsbury, 2012.

Ricks, Thomas E. *The Gamble: General David Petraeus and the American Military Adventure in Iraq 2006–2008.* New York: Penguin Press, 2009.

Rid, Thomas. *Cyber War Will Not Take Place.* New York: Oxford University Press, 2013.

Rousseau, Jean-Jacques. *Discourse on Political Economy and the Social Contract.* Oxford: Oxford University Press, 1994.

Russell, John. *Chechnya – Russia's 'War on Terror.'* New York: Routledge, 2007.

Sapolsky, Harvey M., and Jeremy Shapiro. "Casualties, Technology, and America's Future Wars." *The US Army War College Quarterly: Parameters* 26, no. 2 (1996): 10.

Sassaman, Nathan, and Joe Layden. *Warrior King: The Triumph and Betrayal of an American Commander in Iraq.* New York: St. Martin's Griffin, 2008.

Sauer, Frank, and Niklas Schörnig. "Killer Drones: The 'Silver Bullet' of Democratic Warfare?" *Security Dialogue* 43, no. 4 (2012): 363–80.

Scahill, Jeremy. *Blackwater: The Rise of the World's Most Powerful Mercenary Army*. New York: Nation Books, 2007.

Scahill, Jeremy. *Dirty Wars: The World Is a Battlefield*. New York: Perseus, 2013.

Schleifer, Ron. "Jewish and Contemporary Origins of Israeli *Hasbara*." *Jewish Political Studies Review* 15, no. 1–2 (2003): 123–53.

Schmitt, Michael N. "State-Sponsored Assassination in International and Domestic Law." *Yale Journal of International Law* 17, no. 2 (1992): 609–85.

Schneier, Bruce. *Data and Goliath: The Hidden Battles to Collect Your Data and Control Your World*. New York: W. W. Norton, 2016.

Schulzke, Marcus. "Concealing or Confronting the Costs of War: Modes of Representing Casualties in American News and Entertainment Media." *Critical Studies on Security* (forthcoming).

Schulzke, Marcus. "Ethically Insoluble Dilemmas in War." *Journal of Military Ethics* 12, no. 2 (2013): 95–110.

Schulzke, Marcus. *The Morality of Drone Warfare and the Politics of Regulation*. London: Palgrave, 2017.

Schulzke, Marcus. "The Morality of Remote Warfare: Against the Asymmetry Objection to Remote Weaponry." *Political Studies* 64, no. 1 (2016): 90–105.

Schulzke, Marcus. "The Politics of Attributing Blame for Cyberattacks and the Costs of Uncertainty." *Perspectives on Politics* 16, no. 4 (2018): 954–68.

Schulzke, Marcus. *Pursuing Moral Warfare: Ethical Theory and Practice in Counterinsurgency Operations*. Washington, DC: Georgetown University Press, 2019.

Schulzke, Marcus. "Rethinking Military Virtue Ethics in an Age of Unmanned Weapons." *Journal of Military Ethics* 15, no. 3 (2016): 187–204.

Schulzke, Marcus. "The Unintended Consequences of War: Self-Defense and Violence against Civilians in Ground Combat Operations." *International Studies Perspectives* (forthcoming).

Sellars, Kirsten. *'Crimes Against Peace' and International Law*. Cambridge: Cambridge University Press, 2013.

Shaw, Martin. *The New Western Way of War: Risk-Transfer War and Its Crisis in Iraq*. Malden, MA: Polity Press, 2005.

Shaw, Martin. "Risk-Transfer Militarism, Small Massacres and the Historic Legitimacy of War." *International Relations* 16, no. 3 (2002): 343–60.

Sheldon, John B. "Geopolitics and Cyber Power: Why Geography Still Matters." *American Foreign Policy Interests* 36, no. 5 (2014): 286–93.

Shurkin, Michael. *France's War in Mali: Lessons for an Expeditionary Army*. Santa Monica: RAND, 2014.

Simpson, Emile. *War from the Ground Up*. Oxford: Oxford University Press, 2012.

Singer, P. W. 2008. *Corporate Warriors: The Rise of the Privatized Military Industry*. Ithaca: Cornell University Press.

Smith, Hugh. "What Costs Will Democracies Bear? A Review of Popular Theories of Casualty Aversion." *Armed Forces & Society* 31, no. 4 (2005): 487–512.

Smith, Megan, and James Igoe Walsh. "Do Drone Strikes Degrade Al Qaeda? Evidence from Propaganda Output." *Terrorism and Political Violence* 25, no. 2 (2013): 311–27.

Sorley, Lewis. *A Better War: The Unexamined Victories and Final Tragedy of America's Last Years in Vietnam*. New York: Harcourt Brace & Company, 1999.

Sparrow, Robert. "War without Virtue?" In *Killing by Remote Control*, ed. Bradley Jay Strawser, 84–105. New York: Oxford University Press, 2013.

Spearin, Christopher. "Special Operations Forces a Strategic Resource: Public and Private Divides." *Parameters* (2006): 58–70.

Spruyt, Hendrik. *The Sovereign State and Its Competitors: An Analysis of Systems Change*. Princeton: Princeton University Press, 1996.

Stahl, Roger. *Militainment, Inc.: War, Media, and Popular Culture*. New York: Routledge, 2009.

Stanley, Bruce E. *Outsourcing Security: Private Military Contractors and U.S. Foreign Policy*. Lincoln, NE: Potomac Books, 2015.

Steinhoff, Uwe. "Ethics and Mercenaries." In *New Wars and New Soldiers: Military Ethics in the Contemporary World*, edited by Jessica Wolfendale and Paolo Tripodi, 137–52. Burlington, VT: Ashgate, 2011.

Strawser, Bradley Jay. "Moral Predators: The Duty to Employ Uninhabited Aerial Vehicles." *Journal of Military Ethics* 9, no. 4 (2010): 342–68.

Suchman, Mark. "Managing Legitimacy: Strategic and Institutional Approaches." *Academy of Management Review* 20 (1995): 571–611.

Sumption, Jonathan. *The Hundred Years War, Volume 1: Trial by Battle*. Philadelphia: University of Pennsylvania Press, 1999.

Sun Tzu. *The Art of War*. Los Angeles: Enhanced Media, 2017.

Swofford, Anthony. *Jarhead: A Marine's Chronicle of the Gulf War and Other Battles*. New York: Scribner, 2003.

Tilly, Charles. *Coercion, Capital, and European States, AD 990–1992*. Cambridge, MA: Wiley, 1993.

Thompson, Dennis. "Moral Responsibility of Public Officials: The Problem of Many Hands." *American Political Science Review* 74, no. 4 (1980): 905–16.

Trinquier, Roger. *Modern Warfare: A French View of Counterinsurgency*. Fort Leavenworth, KS: US Army Command and General Staff College, 1985.

Tse-tung, Mao, and Samuel B. Griffith. *On Guerrilla Warfare*. North Chelmsford, MA: Courier Corporation, 2005.

Urban, Mark. *Task Force Black: The Explosive True Story of the Secret Special Forces War in Iraq*. New York: St. Martin's Press, 2010.

US Marine Corps. *US Marine Field Manual on Military Operations in Urbanized Terrain, MCWP 3-35.3*. Washington, DC: Department of the Navy, 1998.

van Baarda, Ted. "The Ethical Challenges of a Complex Security Environment." In *Ethics, Law and Military Operations*, edited by David Whetham, 148–72. London: Palgrave, 2011.

Van Creveld, Martin. *Supplying War: Logistics from Wallenstein to Patton*. Cambridge: Cambridge University Press, 2004.

Van Creveld, Martin. *Technology and War: From 2000 B.C. to the Present*. New York: Macmillan, 1991.

Van Creveld, Martin. *Transformation of War*. New York: Simon & Schuster, 2009.

Vegas, Christopher R. "Fire and Maneuver: Versatility in COIN." *Fires* (October–December 2008): 34–37.

Virilio, Paul. *Desert Screen: War at the Speed of Light*. New York: Continuum, 2002.

Walsh, James Igoe. "Precision Weapons, Civilian Casualties, and Support for the Use of Force." *Political Psychology* 36, no. 5 (2015): 507–23.

Walsh, James Igoe, and Marcus Schulzke. *Drones and Support for the Use of Force*. Ann Arbor: University of Michigan Press, 2018.

Walsh, James Igoe, and Marcus Schulzke. *The Ethics of Drone Strikes: Does Reducing the Cost of Conflict Encourage War?* Carlisle Barracks, PA: Strategic Studies Institute, 2015.

Walzer, Michael. *Arguing About War*. New Haven: Yale University Press, 2004.

Walzer, Michael. *Just and Unjust Wars: A Moral Argument with Historical Illustrations*. New York: Basic Books, 1977.

Wawro, Geoffrey. *The Franco-Prussian War: The German Conquest of France in 1870–1871*. New York: Cambridge University Press, 2005.

Weber, Max. "Politics as a Vocation." Reprinted in *The Vocational Lectures*, ed. David Owen and Tracy Strong. Indianapolis: Hackett, 2004.

Weigley, Russell Frank. *The Age of Battles: The Quest for Decisive Warfare from Breitenfeld to Waterloo*. Bloomington: Indiana University Press, 1991.

Wilcox, Lauren B. *Bodies of Violence: Theorizing Embodied Subjects in International Relations*. Oxford: Oxford University Press, 2015.

Wolford, Scott. *The Politics of Military Coalitions*. Cambridge: Cambridge University Press, 2015.

Wolmar, Christian. *Engines of War: How Wars Were Won and Lost on the Railways*. London: Atlantic Books, 2010.

Woods, Chris. *Sudden Justice: The True Costs of America's Secret Drone War*. London: C. Hurst & Company, 2014.

Woodward, Rachel. "From Military Geography to Militarism's Geographies: Disciplinary Engagements with the Geographies of Militarism and Military Activities." *Progress in Geography* 29, no. 6 (2005).

Zakheim, Dov S. *A Vulcan's Tale: How the Bush Administration Mismanaged the Reconstruction of Afghanistan*. Washington, DC: Brookings Institution Press, 2011.

Zarate, Juan C. *Treasury's War: The Unleashing of a New Era of Financial Warfare*. New York: Public Affairs, 2013.

Zehfuss, Maja. "Targeting: Precision and the Production of Ethics." *European Journal of International Relations* 17, no. 3 (2011): 543–66.

Zetter, Kim. *Countdown to Zero Day: Stuxnet and the Launch of the World's First Digital Weapon*. New York: Crown, 2011.

Index

Printed and bound by CPI Group (UK) Ltd, Croydon, CR0 4YY
16/09/2026
14963487-0007